Organic Field Effect Transistors
Theory, Fabrication and Characterization

Series on Integrated Circuits and Systems

Series Editor: Anantha Chandrakasan
Massachusetts Institute of Technology
Cambridge, Massachusetts

Organic Field Effect Transistors: Theory, Fabrication and Characterization
Ioannis Kymissis
ISBN 978-0-387-92133-4

Embedded Memories for Nano-Scale VLSIs
Kevin Zhang (Ed.)
ISBN 978-0-387-88496-7

Carbon Nanotube Electronics
Ali Javey and Jing Kong (Eds.)
ISBN 978-0-387-36833-7

Wafer Level 3-D ICs Process Technology
Chuan Seng Tan, Ronald J. Gutmann, and L. Rafael Reif (Eds.)
ISBN 978-0-387-76532-7

Adaptive Techniques for Dynamic Processor Optimization: Theory and Practice
Alice Wang and Samuel Naffziger (Eds.)
ISBN 978-0-387-76471-9

mm-Wave Silicon Technology: 60 GHz and Beyond
Ali M. Niknejad and Hossein Hashemi (Eds.)
ISBN 978-0-387-76558-7

Ultra Wideband: Circuits, Transceivers, and Systems
Ranjit Gharpurey and Peter Kinget (Eds.)
ISBN 978-0-387-37238-9

Creating Assertion-Based IP
Harry D. Foster and Adam C. Krolnik
ISBN 978-0-387-36641-8

Design for Manufacturability and Statistical Design: A Constructive Approach
Michael Orshansky, Sani R. Nassif, and Duane Boning
ISBN 978-0-387-30928-6

Low Power Methodology Manual: For System-on-Chip Design
Michael Keating, David Flynn, Rob Aitken, Alan Gibbons, and Kaijian Shi
ISBN 978-0-387-71818-7

Modern Circuit Placement: Best Practices and Results
Gi-Joon Nam and Jason Cong
ISBN 978-0-387-36837-5

CMOS Biotechnology
Hakho Lee, Donhee Ham and Robert M. Westervelt
ISBN 978-0-387-36836-8

SAT-Based Scalable Formal Verification Solutions
Malay Ganai and Aarti Gupta
ISBN 978-0-387-69166-4

Ultra-Low Voltage Nano-Scale Memories
Kiyoo Itoh, Masashi Horiguchi and Hitoshi Tanaka
ISBN 978-0-387-33398-4

Continued after index

Ioannis Kymissis

Organic Field Effect Transistors

Theory, Fabrication and Characterization

Ioannis Kymissis
Columbia University
New York, NY
USA

ISBN 978-1-4419-4711-6 e-ISBN 978-0-387-92134-1

Printed on acid-free paper.

springer.com

To my wife, Carisa.

Preface

I have several people to thank for making this book possible. I have learned a great deal from the individuals I have worked with, starting with my mentor at IBM Research, Christos Dimitrakopoulos, who introduced me to the field and taught me virtually everything I know. I also have to also thank Sam Purushothaman, Tayo Akinwande, Vladimir Bulović, and Charlie Sodini, for their guidance and mentorship.

I have had the honor of working with several students on OFET-related projects, including Annie Wang, Kevin Ryu (who also provided me with a number of figures in the text), Ivan Nausieda, Yu-Jen Hsu, and Zhang Jia. There is no question that I have learned more from them than they have from me. In addition to the students I have worked with, Dennis Ward was also instrumental in developing and fine-tuning the process presented in appendix B.

The input from several colleagues was particularly valuable for this manuscript, and I'd like to thank Charlie Sodini and Marshall Cox for their time and comments. I especially have to thank Fabiola Soong, who reviewed several versions of the manuscript and provided valuable feedback on both the technical content and style of the book.

I hope that this book serves as guide and reference to the field and I welcome your comments, criticisms, and suggestions at `johnkym@ee.columbia.edu`.

Ioannis Kymissis,

New York City,
August 2008

Contents

1 **Introduction** . . . 1
1.1 Why OFETs? . . . 1
1.2 A very brief history of OFETs . . . 2
1.3 Organization of this book . . . 3

2 **The physics of organic semiconductors** . . . 5
2.1 Free electron model . . . 5
2.1.1 Carbon is special . . . 5
2.1.2 Conjugated molecules as a particle-in-a-box . . . 5
2.1.3 Energy levels in semiconducting polymers . . . 10
2.1.4 Applying the free electron model to small conjugated molecules . . . 11
2.2 Charge and energy carriers in conjugated molecules . . . 11
2.2.1 Carriers in organic semiconductors: n-type or p-type? . . 14
2.2.2 Electron rich and electron poor materials . . . 15
2.3 Conclusion . . . 15

3 **Organic semiconductor materials for OFETs** . . . 17
3.1 Major classes of organic semiconductors . . . 17
3.1.1 Polymer semiconductors . . . 17
3.2 Small molecule semiconductors . . . 20
3.2.1 Air stability . . . 25
3.2.2 Organic conductors . . . 26
3.3 Conclusions . . . 26

4 **Basic OFET fabrication** . . . 29
4.1 Introduction . . . 29
4.2 Basic OFET structure and operation . . . 29
4.3 Unit operations . . . 30
4.3.1 Thermal evaporation . . . 30
4.3.2 Liquid deposition . . . 40

4.3.3 Polymer CVD ... 41
4.3.4 Other applicable PVD and CVD processes ... 42
4.3.5 Subtractive patterning operations ... 43
4.3.6 Etching ... 45
4.4 Processing considerations for high crystallinity ... 45
4.4.1 Polymer crystallinity ... 45
4.4.2 Stranski-Krastanov growth of small molecule crystallites 46
4.4.3 Threshold voltage and bias stress ... 49
4.5 Several archetypical process flows and variants ... 49
4.5.1 Shadow masking ... 50
4.5.2 Parylene encapsulation ... 51
4.5.3 PVA resist ... 52
4.5.4 Subtractive inkjet/digital lithography ... 53
4.6 Conclusions ... 55

5 Advanced OFET fabrication ... 57
5.1 Introduction ... 57
5.2 Source and drain contacts ... 57
5.2.1 Work function considerations ... 57
5.2.2 Top vs. bottom contacts ... 60
5.2.3 Treatment of contacts ... 61
5.2.4 Creation of lithographic top contact devices ... 63
5.3 Gate dielectrics ... 63
5.3.1 Characteristics of gate dielectrics ... 63
5.3.2 Crystal structure improvement ... 64
5.3.3 SAM gate dielectrics ... 65
5.3.4 Introduction of surface dipoles ... 65
5.3.5 Functional gate dielectrics ... 66
5.4 Air sensitivity and encapsulation ... 66
5.5 Emerging deposition and patterning processes ... 68
5.5.1 LITI ... 68
5.5.2 OVPD ... 68
5.5.3 Surface energy modulation ... 68
5.6 Alternative OFET designs ... 69
5.6.1 SIT ... 69
5.6.2 Reduced patterning processes ... 70
5.6.3 Electrochemical OFETs ... 72
5.7 Self-aligned OFETs ... 73
5.8 Conclusions ... 74

6 Modeling and characterization ... 75
6.1 Models ... 75
6.1.1 The role of models ... 75
6.1.2 The IEEE 1620 standard ... 75
6.1.3 Long channel silicon device operation ... 75

6.1.4 Long channel silicon device model ... 78
6.2 Parameters ... 80
6.2.1 Mobility ... 80
6.2.2 Threshold voltage ... 82
6.2.3 Contact resistance ... 86
6.2.4 Hysteresis/bias-stress ... 88
6.2.5 Gate leakage ... 90
6.2.6 Subthreshold slope ... 91
6.2.7 Output conductance ... 91
6.3 Characterization ... 92
6.3.1 Gate sweep/transfer characteristic ... 93
6.3.2 Drain sweep/output characteristic ... 93
6.3.3 Capacitance ... 94
6.3.4 Gate leakage ... 95
6.4 Device model ... 95
6.5 Parameter summary ... 96
6.5.1 The limits of curve fitting in amorphous systems ... 98
6.5.2 Measurement and reporting ... 101
6.6 Conclusions ... 101

7 OFET applications ... 103
7.1 Displays ... 103
7.2 Mechanical sensors ... 105
7.3 Imagers ... 106
7.4 RFID and logic ... 107
7.5 Conclusions ... 107

A Appendix A: Review articles ... 109

B Appendix B: Recipes and equipment ... 111
B.1 Overall process flow ... 111
B.2 Notes on equipment ... 111
B.3 Substrate preparation ... 112
B.4 Gate deposition ... 113
B.5 Photolithography ... 114
B.6 Gate dielectric ... 114
B.7 Source/drain ... 115
B.8 Semiconductor layer ... 115
B.9 SAM treatment ... 116
B.10 Process summary ... 116

C Appendix C: Device layouts ... 119
C.1 Design rules ... 119
C.1.1 Pattern and overlay parameters ... 119
C.1.2 Test mask for pattern and overlay parameters ... 119

C.1.3 Determining design rules 121
C.2 Alignment marks 122
C.3 Transistor layout 123
C.3.1 Probe/bond pads 123
C.3.2 Channel layout 124
C.3.3 Circuit layout 125
C.4 Transfer line 125
C.5 Test structures 126
C.5.1 Resistivity 127
C.5.2 Capacitance 127
C.5.3 Optical thickness 127
C.6 Quad mask design 128

D Appendix D: Logic gates 129
D.1 Inverters 129
D.2 NAND and NOR gates 130
D.3 Ring oscillators 131
D.4 SRAM 133
D.5 Flip flops and code generators 134
D.6 Complementary OFET circuits 136

References 137

Index 147

1

Introduction

1.1 Why OFETs?

Organic field effect transistors (OFETs) have grown from a curiosity to a significant research area over the past two decades.

Most of the interest in OFETs stems from the low thermal budget required to fabricate these devices and their high degree of mechanical flexibility. These characteristics follow from two basic properties of organic semiconductors:

1. organic semiconductors enjoy semiconducting properties on the molecular level and
2. bonding between organic molecules can be weak, usually formed by van der Waals bonds

Because no covalent bonds need to be broken or re-formed, organic semiconductor materials can be manipulated using a small energetic input. Printed, evaporated, and ablated materials can form semiconducting channels on virtually any substrate. No epitaxial templating is required to achieve at least some semiconducting properties, and materials can be engineered to self assemble into favorable configurations for device performance in low thermal input processes since that assembly is not energetically expensive.

This weak bonding and ease of manipulation often comes at some price in performance. Through improvements in materials, processing, and circuit design, however, performance which by many metrics exceeds that of amorphous silicon can be achieved at a thermal budget unmatched by most inorganic semiconductors (some recently developed metal oxides excepted). This allows OFETs to be placed on virtually any substrate or surface–plastic foils, on other active devices, or glass. A large number of applications have been proposed which take advantage of the simple deposition and patterning processes, ability to integrate with other devices, and the mechanical flexibility that the weakly bonded thin film format offers, or combinations of these properties are under development.

1.2 A very brief history of OFETs

The metal-oxide-semiconductor capacitor using organic semiconductors was demonstrated by Ebisawa, Kurokawa, and Nara at NTT in 1982. The device was fabricated using polyacetylene as the semiconductor on a polysiloxane gate dielectric and used aluminum for the gate and gold for the source and drain electrodes. While the device operated in depletion mode and showed only a few percent current modulation when analyzed for transconductance, the authors clearly recognized the concept's potential and ended the paper suggesting that this type of device "appears promising" for thin film transistors [1].

The next significant milestone was the development of the first organic field effect transistor (OFET) with recognizable current gain by an in-situ polymerized polythipohene transistor by Tsumura, Koezuka, and Ando of Mitsubishi Chemical in 1986 [2].

Interest in organic semiconductors began to intensify in the late 80's and early 90's, in large part because of reports describing OFETs, an organic heterojunction organic solar cell by Tang in 1986 [3], and organic light emitting diode by Tang and Van Slyke in 1987 [4]. A number of programs were established at academic and industrial research centers to investigate one or more of these device architectures.

Many of the improvements in OFET technology have followed improvements in the available materials for device fabrication, led by synthetic chemists who understand the needs of electrical devices. The first OFET devices were fabricated by directly polymerizing insoluble films of polyacetylene and polythiophene onto substrates, which is a technique that is limited in its large area applicability. A soluble form of polythiophene, developed by Jen et al. [5], and applied to OFETs by Assadi, et al. [6] ignited excitement about the possibility of printable semiconductor systems which could be made with the same economies of scale as printed paper media. Soluble polymers, dispersible polymers, and a range of small molecule systems have been developed for compatibility with a wide range of deposition and patterning technologies. Specialized materials for electrodes and insulators also continue to be developed and refined.

Concurrent with this improvement in materials has been improvement in fabrication techniques and characterization. Better processing has contributed to improved device performance, reduced deposition cost and time, and to the development of architectures and process flows suitable for target applications. Better characterization has led to an improved understanding of device operation and performance and a metric for materials diagnosis.

OFETs stand at a crossroad today. Performance has, by many metrics, exceeded that of amorphous silicon. Several industrially scalable processing strategies have been developed, and it is quite likely that the cost and energetic input of such processes is significantly less than that incurred for amorphous silicon. Industrial acceptance, however, has been limited. It remains one of the primary goals of the OFET research community to develop applications which uniquely exploit the properties of OFETs and translate them into killer applications which will cement OFETs' relevance and exploit their unique characteristics.

1.3 Organization of this book

This book is intended as an introduction and reference to OFETs. It discusses the fundamental mechanisms at play in their fabrication, their operation, and their characterization. It also discusses many of the practical issues associated with OFET fabrication, characterization, modeling, and integration with other devices.

Chapter 2 provides a summary of the chemical physics of organic semiconductor operation. It explains why carbon is so special and how its unique properties lend it to the formation of an unusual class of semiconductor materials.

Chapter 3 explains the major classes of OFET materials which are in use, their properties, and some of the advantages and disadvantages of each system. It also discusses many of the parameters which go into material design and selection of semiconductors and doped organic conductors which can be used as electrodes.

Basic processing of OFETs is explained in Chapter 4. It begins with an explanation of the foundational unit operations for OFET fabrication and walks through several archetypical process flows which make OFET integrated circuits using these techniques. Chapter 5 continues this discussion for more advanced techniques. Several strategies for increasing throughput and performance are discussed here.

Modeling and device characterization is explained in chapter 6. The legacy strategy usually encountered in the literature, the IEEE standard method, and several emerging strategies for device characterization are presented. This section also discusses a number of non-ideal effects and builds a small signal device model which can be used for characterization and modeling purposes. Chapter 7 summarizes several application areas in which OFETs have been applied and shows the structure of the circuits used in many of these applications.

The appendices provide some additional material which can provide a starting point for further research into OFETs. While some individual papers are referenced in the text, many review articles provide greater detail and more references for specific topics; several review papers are listed in Appendix A. Appendix B provides detailed recipes for a transistor fabrication process flow. Mask layout and design is discussed in Appendix C.

2

The physics of organic semiconductors

This chapter presents a simple introduction to the chemical physics of organic semiconductors, which is a rich and complicated topic. The hope is to develop some intuition which links chemical structure to optical properties and electronic behavior. A much more through discussion is presented is a number of chemical physics-oriented texts including [7] and [8].

2.1 Free electron model

2.1.1 Carbon is special

Carbon is unusually rich in its chemical behavior for a number of reasons:

- Carbon has a relatively small size. This reduces the steric hindrance in carbon-containing molecules, and allows a larger variety of compounds to be created.
- Carbon has a moderate electronegativity. This allows it to form covalent bonds with virtually all materials including other carbon atoms.
- Carbon is a group IV material. This allows it to form four bonds, increasing its chemical versatility and helping it form long chain polymers.
- Carbon hybridizes in a number of forms. This allows for a wide variety of bonding configurations including single, double, and triple bonds as well as a range of resonance forms.

It is because of these properties that both synthetic organic chemistry and natural synthetic processes are able to deliver a wide range of carbon-based materials with tailored mechanical, optical, electrical, catalytic, and other properties from substantially similar building blocks.

2.1.2 Conjugated molecules as a particle-in-a-box

A covalently bonded molecule is a relatively fixed arrangement of atoms. The chemical character of a molecule is determined by its electronic structure and the ease with which that electronic structure can be rearranged to form new bonds.

The atomic orbitals in carbon have the ability to hybridize into a number of geometries. VESPR theory and traditional orbital modeling show that three hybrid orbitals of relevance for carbon are sp, sp^2, and sp^3. These are formed by the superpositon of s and p orbitals; calculations of the electron density probability surfaces forming individual sp^2 orbitals are shown in Fig. 2.1 (c).

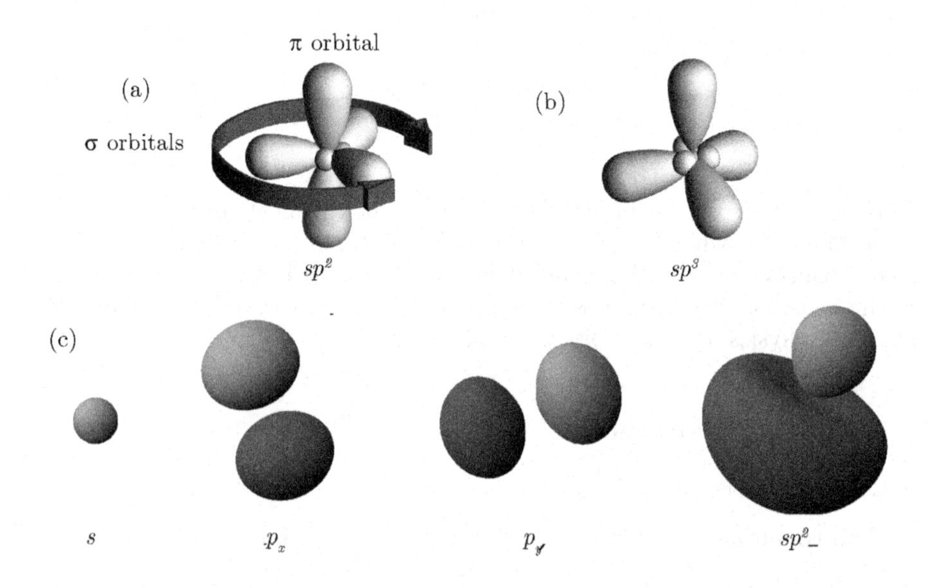

Fig. 2.1. (a) and (b) schematically show the geometry of the sp^2, and sp^3 orbital structures. (c) shows a calculated constant electron density probability surface for the constituent components of the sp^2 orbital (one s and two of the three p orbitals) as well as one lobe of the hybrid orbital itself.

The graphic in Fig. 2.2 shows the chemical structure and a three dimensional model of a molecule (ethylene) that consists of a double bonded pair of carbon atoms which are both sp^2 hybridized. Note that the bonding structure is the same trigonal planar structure shown in Fig. 2.1 (b), with bond angles close to 120° (within a degree in the case of ethylene).

Ethylene has two carbon-carbon bonds, schematically shown in 2.2 (c). One can be attributed to the hybridized sp^2 orbital, and the second is the sharing of the remaining unhybridized p_z orbital. The sp^2 bond is known as a sigma (or σ) bond, and the p_z bond is known as a pi (or π) bond. Single bonds among sp^2 hybridized molecules are formed using sigma (hybridized) orbitals, double bonds are typically a product of one a sigma and one pi bond.

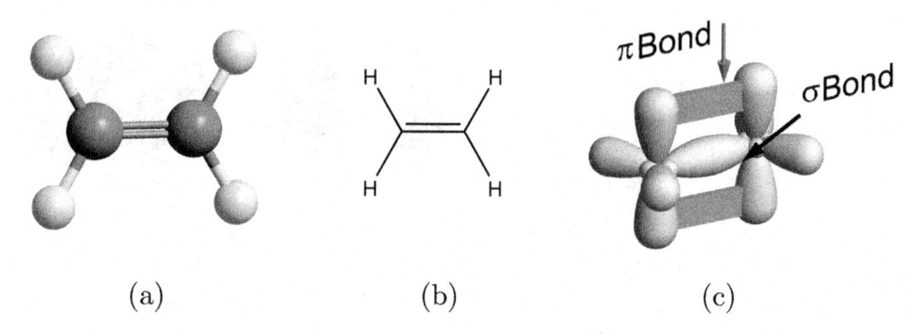

Fig. 2.2. Several representations of ethylene, which has two sp^2 bonded carbon atoms. (a) is a ball and stick model, (b) shows a chemical structure, and (c) is a schematic orbital diagram. Note the trigonal planar arrangement of the bonds between the atoms–this arrangement is the consequence of the hybrid orbital structure.

As molecules increases in size, their electron density becomes more complicated. There is some ambiguity in the exact configuration of the bonding structure if there are both single and double bonds present in a molecule. If one takes benzene, for example, there are two familiar resonance forms, shown in Fig. 2.3. In both of these resonance forms, a single bond (which is the sigma bond) always remains between the carbon atoms. The pi bond is able to move from atomic site to atomic site and is effectively delocalized among the atoms in the molecule.

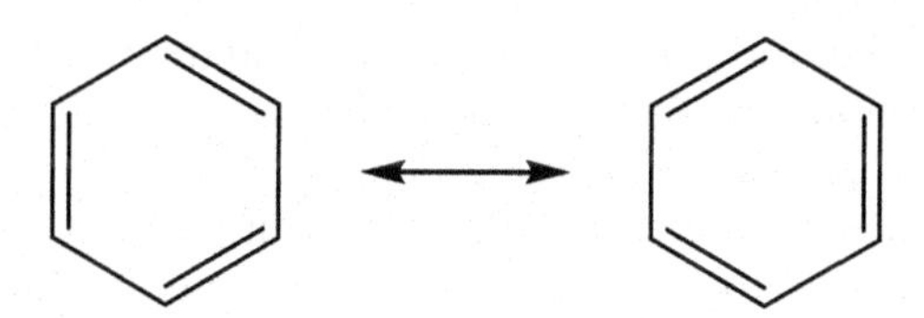

Fig. 2.3. Two resonance forms of benzene.

The alternating single and double bonds seen in benzene are an indication that the carbon atoms in the molecule are sp^2 hybridized. Materials in which neighboring carbon atoms are sp^2 hybridized, forming delocalized clouds of pi electrons, are said to be conjugated. The electron density model of benzene shown in Fig. 2.4 (a) illustrates this delocalization.

Benzene is a relatively simple and somewhat unusual example; there are two clearly defined resonance forms and they are equivalent in energy. If we look at larger molecules, there are many more resonance forms which lack

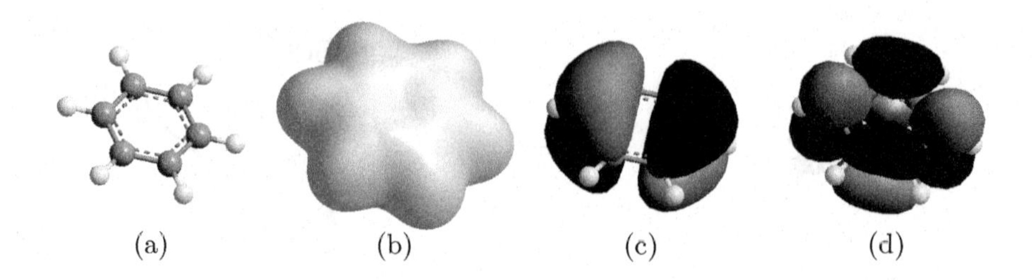

Fig. 2.4. Four views of benzene. (a) a chemical model of the molecule; (b) an isocharge surface; (c) the HOMO, or highest occupied molecular orbital; (d) the LUMO, or lowest unoccupied molecular orbital.

the symmetry of benzene and some will be more energetically favorable than others.

Resonance is not a digital phenomenon in which the bonding jumps from one discrete state to another; it is the representation of a superposition of many states in the limited pictorial model of the chemical structure. Put differently, the electrons which are present in these superimposed states are not locked into a particular rigid configuration and instead enjoy a greater or lesser degree of delocalization.

This pi electron delocalization forms an energy well situation; the electrons are the particles, and the molecule forms a box. The energy levels of this configuration are determined by the occupancy and extent of the box and can be modeled, to first order, using the traditional particle in a box formulation.

On an isolated molecule, these energy levels are discrete and known as molecular orbitals. They are the product of the combination of atomic orbitals that comprise the molecule. Some of the molecular orbitals will be filled with electrons when the molecule is in the ground state, and higher levels will be empty. The two most interesting molecular orbitals are the frontier states: the highest occupied molecular orbital (HOMO) and the lowest unoccupied molecular orbital (LUMO). The creation and absorption of packets of energy (including photons) is governed by transitions between empty and full orbital states.

Generally speaking, the larger the conjugated network is on a molecule, the smaller the HOMO-LUMO gap will be because of the increased extent of the potential well. If one looks at the series of molecules in the same series as butadiene there is a clear trend in the HOMO-LUMO gap with increasing chain length.

In an organic semiconductor solid the collective energy transitions are not as sharp as this picture suggests; the interaction between molecules spreads out the energy levels due to a variety of intermolecular interactions. Discrete levels can be observed in molecules while they are isolated in the gas phase (and therefore in a uniform environment).

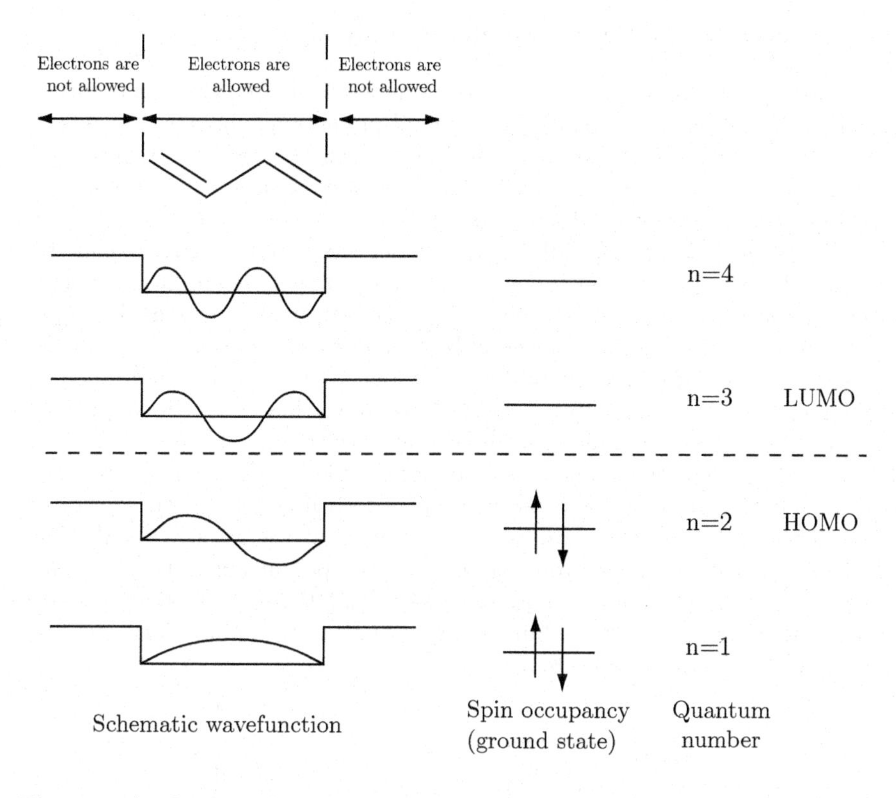

Fig. 2.5. A schematic diagram showing the formulation of the energy structure of butadiene as a particle in a box problem. Only the pi electrons are shown. In the free pi-electron model (which is a simplification) each state can be occupied by two carriers, each with opposite spin.

Table 2.1. The HOMO-LUMO gap for a series of short conjugated molecules, calculated using an Extended Huckel model. Longer chains form a larger box, and therefore have a smaller separation between energy levels.

Number of C atoms	HOMO-LUMO gap
4	10.56eV
6	8.634eV
8	6.054eV
10	4.986eV

2.1.3 Energy levels in semiconducting polymers

The view of holes and electrons pi-orbital molecule as free carriers is rather simplistic. It allows us to make some general predictions about the trends in energy levels if we use it carefully, but does not extend well to polymers.

One would expect that in an infinite sp^2 bonded system, such as a conjugated polymer, there would be no bandgap.

Infinite one dimensional conjugated polymers still maintain a finite bandgap due to a dimerization phenomonon known as the Pierels instability. The free electron approximation (which views the pi electron region as an isotropic Drude electron gas) does not adequately explain this scenario.

The Pierels instability guarantees that one dimensional metals are not stable, and that an alternative state is always lower in energy. In conjugated organic materials this manifests itself as a dimerization involving two unequal bond strengths (to view these as single and double bonds is not entirely accurate, but this is a common picture). The derivation of the energy levels is straightforward [9]; the establishment of unequal energy levels and what is effectively a periodic electron potential on the polymer maintains a finite bandgap even in an infinitely long molecule. That HOMO-LUMO gap's size is determined by the degree of dimerization and the recombination length of carriers in the conjugated system.

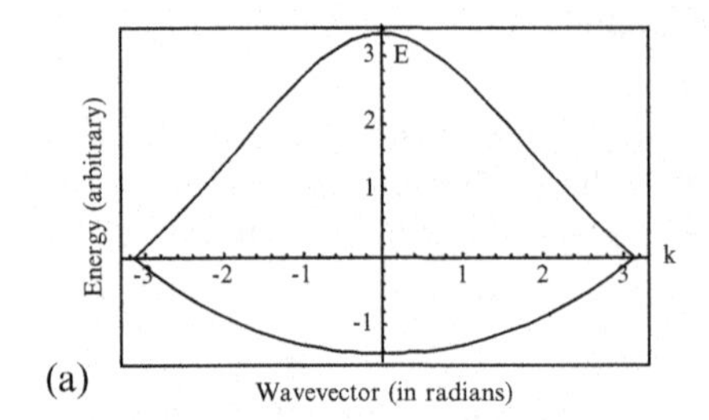

(a)

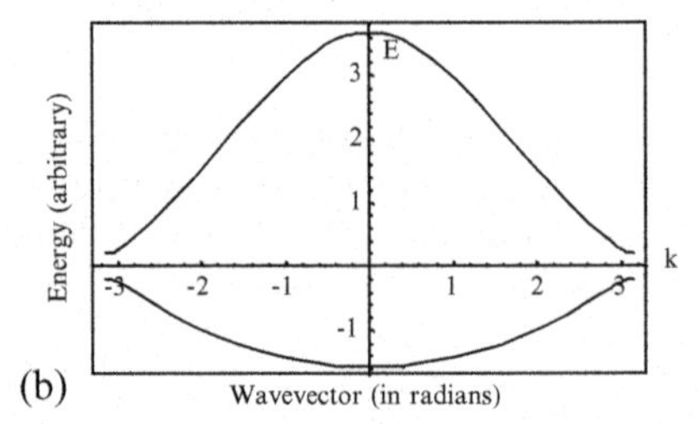

(b)

Fig. 2.6. The energy level diagram for (a) an unphysical infinite one dimensional metallic version of polyacetylene and (b) the more realistic dimerized polyacetylene molecule. The monomeric uniform molecule does not exhibit a bandgap because of its infinite size. A HOMO-LUMO gap exists despite the infinite extent of the molecule because of the Pierels instability which dimerizes the molecule and separates the bands at the Brilloin zone edge. These dispersion relationships are calculated using the derivation presented in [9] and [10].

Higher dimensional pi-electron carrier systems, like carbon nanotubes and graphene, are not subject to the Pierels instability and can be metallic or semi-metallic.

2.1.4 Applying the free electron model to small conjugated molecules

A relatively accurate determination of the energy levels in small molecules can be determined in a manner similar to that employed above for polyacetylene. Tight binding methods using more sophisticated Hamiltonians tend to give the best results. It is also possible to estimate the energy levels in small molecule semiconductors using a free electron model. In this approximation, the length of the conjugated portion of the molecule determines the energy levels [7]. For acenes, which are cyclic conjugated molecules, the Perimeter Free Electron Orbital model (PFEO) can be used.

The PFEO model is a restatement of the free electron model presented above, but assumes that the bounding container is the perimeter of the molecule. The first four levels expected from the wavefunction are shown in Fig. 2.7.

The energy levels can then be calculated from the frontier orbitals. Sherr has compiled the following results [11]:

Table 2.2. Predicted and measured HOMO-LUMO transitions for several acenes, PFEO and experimental values are from Sherr [11].

Material	PFEO	Experimental value
Benzene	6.3eV	6.0eV
Napthalene	3.8eV	4.3eV
Anthracene	2.7eV	3.3eV
Tetracene	2.1eV	2.6eV
Pentacene	1.7eV	2.1eV

While not completely accurate, the PFEO model provides an estimate of the first level excitation energy and explains the trend in HOMO-LUMO gap. As the molecule increases in size the HOMO-LUMO gap decreases in magnitude and it becomes easier to generate an excited state. Technologically this has some ramifications; for example, hexacene's very small HOMO-LUMO gap makes its electrons easy to excite. The material is easily oxidized in air and unsuitable for air stable transistors. The HOMO and LUMO surfaces for pentacene are shown in Fig. 2.8. It is clear that the wavefunction indeed travels around the perimeter of the molecule and this distance is a significant determiner of the electronic properties of the material.

2.2 Charge and energy carriers in conjugated molecules

If we look carefully at the resonance forms of conjugated molecules we can also explain the formation of charge carriers on organic semiconductors. If we

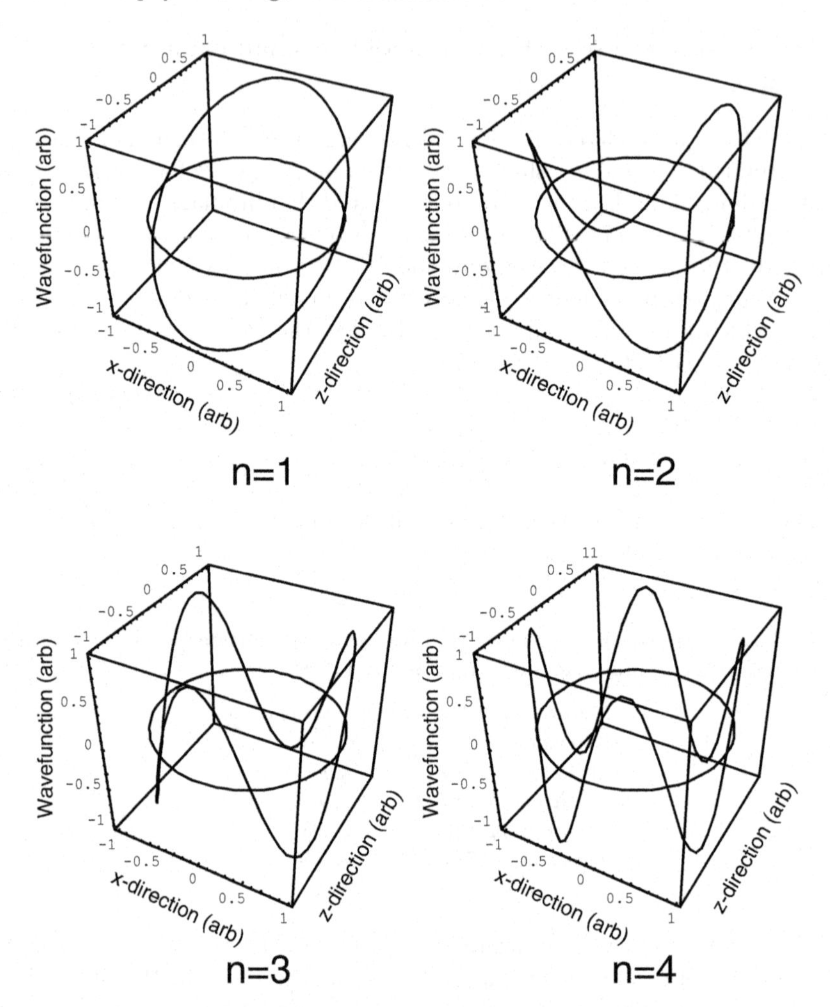

Fig. 2.7. A representative schematic of the first four PFEO wavefunctions

look at polyacetylene's structure, for example, shown in Fig. 2.9, we see two states which are energetically degenerate.

It is possible for these two states to coexist on the same chain with a boundary between them. The entropy of a perfect chain is low, and one can imagine a certain density of these defects existing even with only a small thermal excitation. At the phase wall between these two states, we might see a charge carried on the polymer chain, schematically illustrated in Fig. 2.10.

This combination of two resonance forms is known as a polaron. Charge can sustain itself on the unsatisfied bond indicated by the + sign in Fig. 2.10, and this charge can be moved on the molecule by an electric field. This phe-

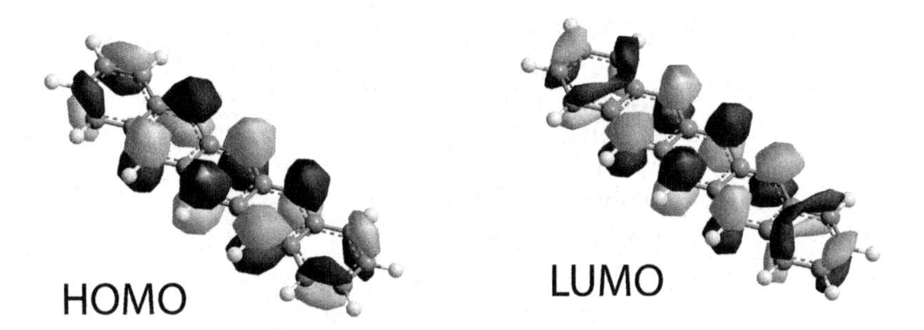

Fig. 2.8. The HOMO and LUMO of pentacene.

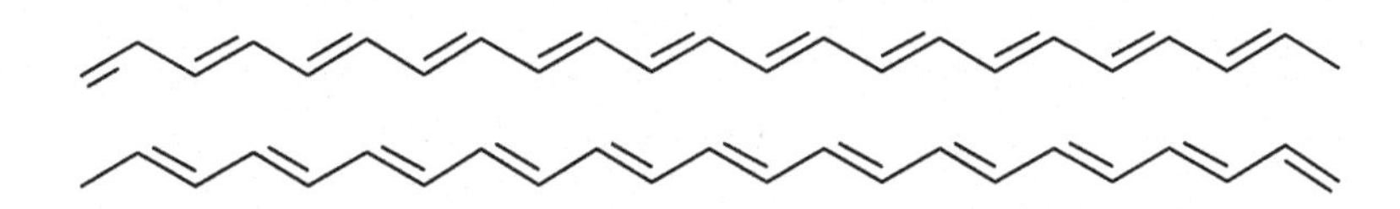

Fig. 2.9. Two resonance forms of polyacetylene.

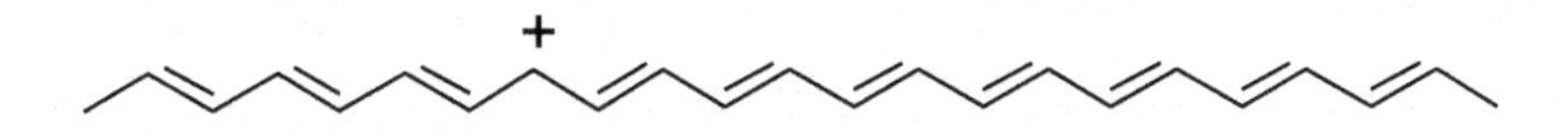

Fig. 2.10. A single positively charged polaron on polyacetylene.

nomenon is responsible for the generation charge states in organic semiconductors. A similar situation can be generated by a dopant which might oxidize or reduce one of the bonds along the chain creating a free electron or hole. Polarons can also become combined in more complicated structures with greater charge or carriers such as excitons where positively and negatively charged polarons attract each other and form an electrically neutral energy carrier. The ability for polarons to carry spin is also determined by their occupancy.

A number of charge and energy carriers can be found in small molecule and polymer organic semiconductors, a few are summarized in table 2.2 [12].

Virtually all polymeric organic semiconductors have a conjugated portion with the same sp^2 configuration as polyacetylene. Fig. 2.11 shows three representative materials; each of the three is a structure grafted onto a polyacetylene backbone.

The situation is similar in oligomeric semiconductors. The delocalization of electrons in the pi electron cloud decreases the energetic penalty for the manipulation of charge density from the ionization energy of carbon (which is an almost insurmountable 11eV or so) to a small fraction of that energy;

Table 2.3. Several possible charge and energy carriers on organic semiconductors.

Carrier	Components
Negative bipolaron	- -
Negative polaron	-
Exciton	+ -
Positive polaron	+
Positive bipolaron	+ +

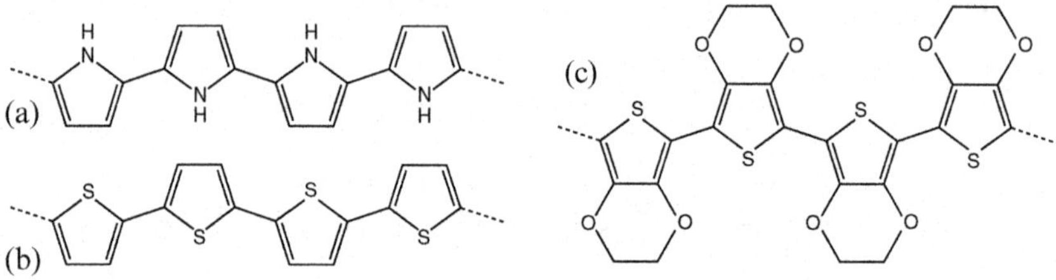

Fig. 2.11. Three organic semiconducting polymers. (a) Polypyrrole, (b) plythipohene, and (c) poly(3,4-ethylenedioxythiophene) (PEDOT). All three have an sp^2 conjugated backbone identical to polyacetylene. It is along the pi-bonds in this backbone that these materials exhibit carrier delocalization and electrical activity.

one that can be achieved with chemical doping or the application of relatively small electric fields.

2.2.1 Carriers in organic semiconductors: n-type or p-type?

The transport of charge in organic semiconductors typically follows a hopping mechanism and is governed by the class of transport models used to model highly disordered systems (there is a further discussion in Chapter 6). The density of states of these disordered systems is often asymmetrical and presents a significantly larger barrier to the formation of one or another type of carrier. Many amorphous or polycrystalline organic semiconductors exclusively support positive or negative charge carriers, but not both.

In crystalline semiconductor technologies, n-type and p-type refer to the type of dopant, and therefore majority carrier, in a semiconductor. In crystalline materials both holes and electrons can usually be transported reasonably well. Disordered electronic systems, on the other hand, often only support or strongly favor one type of charge carrier and are more properly referred to as hole or electron transporting. It is, nevertheless, common in the literature to refer to hole transporting disordered semiconductor materials as p-type and electron transporting materials as n-type because this describes their majority carriers and semiconducting behavior.

This topic has become of greater interest now that some experiments suggest that a symmetrical density of states (and ambipolar behavior as a consequence) may be more general than previously thought and that small levels of unintentional doping from electronegative groups on the gate dielectric cause traps that contribute to the observed asymmetry in the density of states [13].

2.2.2 Electron rich and electron poor materials

If we look at a popular organic semiconductor, say pentacene, the hydrogen atoms which surround the carbon backbone are less electronegative than the carbon backbone itself and lend some electron density to the delocalized pi-electron cloud. The electron rich conjugated molecule has difficulty accepting another electron, but is able to lose an electron with relative ease. As a consequence, positive charge carriers dominate transport in pentacene thin films.

A large number of electron transporting organic semiconductors have also been synthesized; these incorporate electron withdrawing groups which make an electron deficient core which carries electrons more readily.

It is possible to significantly shift the density of states and the carrier type by reacting organic semiconductors with strongly electropositive or electronegative materials such as alkali metals (e.g. K) and halogens (e.g. I) [14] [15]. Doping with small atomic species is typically reversible, however, and this strategy has not been especially effective in creating devices with stable characteristics. The use of larger organic dopants, which are unable to diffuse, has been demonstrated to be effective in stably doping organic semiconductors [16].

2.3 Conclusion

Organic field effect transistors have unique material properties which both make them attractive as a material system and somewhat complicated to handle. Understanding the basic chemical physics that underlies device operation, energy levels, self-organization, and stability informs allows for the rational design of new materials, processes, and devices.

A difficult question is which organic semiconductor material is best to start with for a device application. There is no single general answer, but several considerations should be made including:

- What mobility, threshold voltage, and on/off ratio are required?
- What materials are commercially available in a suitable form (e.g. in solution, powder, etc.) and purity?
- Who will purify the material? Is it possible to purify internally (or have the material of interest purified by the vendor or a third party)?
- Is this material compatible with the desired electrodes and gate dielectric?
- Is it important for the material to lack (or have) a strong photocurrent response or fluorescence?

- Is the material stable, and are facilities available for handling and packaging non air-stable materials?
- For commercial applications, are there regulatory or IP obstacles which need to be overcome to legally and safely use the material in the target manufacturing environment and target markets of interest?

3

Organic semiconductor materials for OFETs

3.1 Major classes of organic semiconductors

It is almost impossible to catalog all OFET materials; new materials and forms of known materials are being developed as processing and synthetic techniques improve. Most materials in common use can, however, be placed into one of a few classes. A more thorough delineation of semiconducting materials for OFETs and emerging material classes can be found in several excellent review articles, some of which are listed in Appendix A.

Organic semiconductors are generally small molecules or polymers made from a small foundational group of conjugated monomer units, shown in Fig. 3.1. How these elements are connected has a profound influence on how they are processed and how they perform.

3.1.1 Polymer semiconductors

The following sections discuss a few model polymer systems and the structure-function relationships which make them suitable for use on OFETs. Many other polymers have been developed and demonstrated in OFETs; the principles which make advanced polythiophene and polyfluorene polymers and their copolymers attractive can be applied to many of these other systems as well.

Regioregular polythiophenes-a model compound

Polythiophenes are an excellent example of how structure engineering can improve the performance of a material family [17]. Initial work in polythiophenes as transistor materials was started in 1986 and used electropolymerized polythiophenes (first used as conductors [18]) which were electrochemically dedoped to produce a material which could be accumulated and depleted [2]. Polythiophene is insoluble and electropolymerization requires electrical access to all points, which complicates deposition in circuit forming devices. By

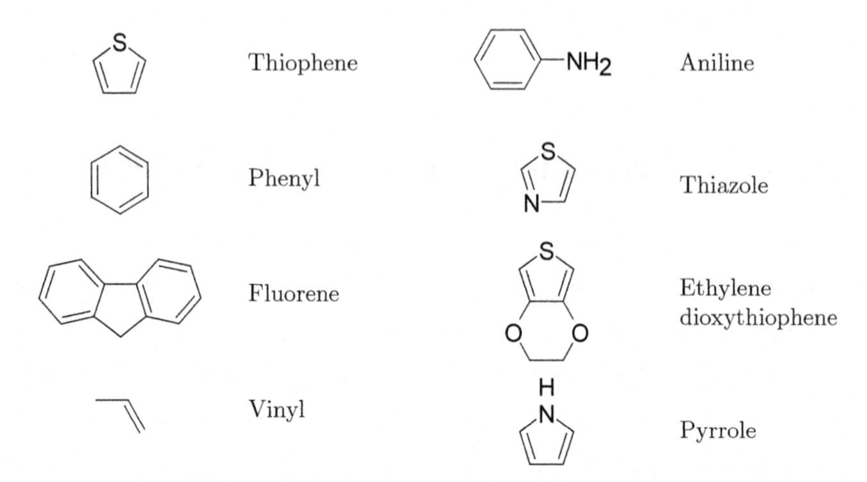

Fig. 3.1. Some of the more common repeating functional units in conjugated organic materials. Most organic semiconductors and conductors are made from fused or linked elements like these, which are rich in sp^2 hybridized carbon atoms and delocalized pi electrons. It should be noted that chemical synthesis of organic semiconductors and conductors is often not performed using these materials as starting ingredients.

adding alkyl side chains, a polythiophene derivative was developed by Jen, et al. [5] which adds alkyl side chains to the thiophene monomer and makes the polymer soluble without disurbing conjugation in the backbone. Assadi et al. applied this more practical material to make transistors [6], which were significantly easier to process while showing essentially the same performance as the electrochemically polymerized material.

Devices made using these two techniques were essentially amorphous. Electropolymerized polythiophene has no strong organizing force for crystallinity, and the large groups used to solubilize polythiophene sterically interfere with chain-to-chain packing.

A significant breakthrough was made in 1995 by Chen et al. with the development of regiocontrolled polythiophene using a specialized zinc catalyst [19]. This regioregular alkyl substituted polythiophene exhibited significantly improved behavior due to superior packing while retaining the processing advantages of solubility. Subsequent refinement in increasing the regioregularity and purity of the material have produced progressively increasing gains in performance, now rivaling that of amorphous silicon.

Polythiophene based liquid crystalline materials

The organizational capabilities of soluble polythiophenes can be further enhanced by creating structures which retain a high degree of regioregularity

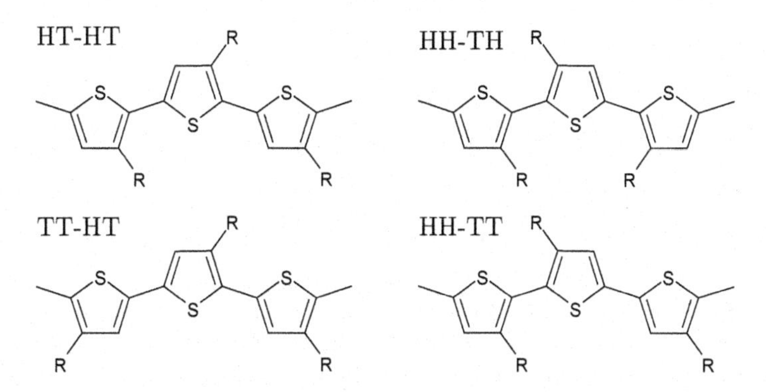

Fig. 3.2. The four possible conformational triads of alkyl substituted polythiophene, where R is the alkyl functionalizing group [20]. The HT-HT conformation has been found to stack the best. These triads can be monitored using NMR spectroscopy to determine the degree of regioregularity in a sample.

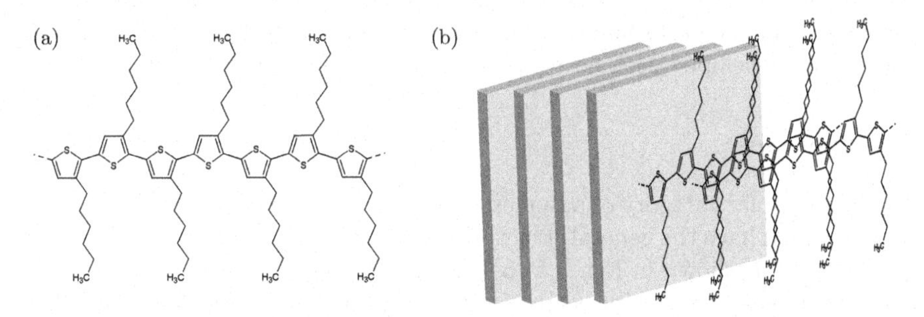

Fig. 3.3. (a) The structure of regioregular HT-HT P3HT and (b) a schematic showing its lamellar stacking structure.

but have a reduced activation energy for conformational readjustment and crystallization. One approach to achieving this is through reduction of the steric hindrance caused by the solubilizing side groups of the polymer material. A family of polythiophenes was developed by Ong, et al. [21] in which a 4 unit thiophene repeating unit is used, only two of which are substituted (Fig. 3.4 (a)). Due to its greater orientational freedom and decreased crowding between chains this material is liquid crystalline and can be thermally processed into highly crystalline thin films by controlled heating and quenching of the crystalline state. The reduced substitution also leads to greater oxidative stability.

Further optimization can be achieved by engineering the monomer units to increase the pi-pi coupling between adjacent units. A copolymer of thienothiophene and thiophene repeat units (Fig. 3.4 (b)), for example, has been shown to stiffen the polymer backbone. This increases the crystallinity of the device

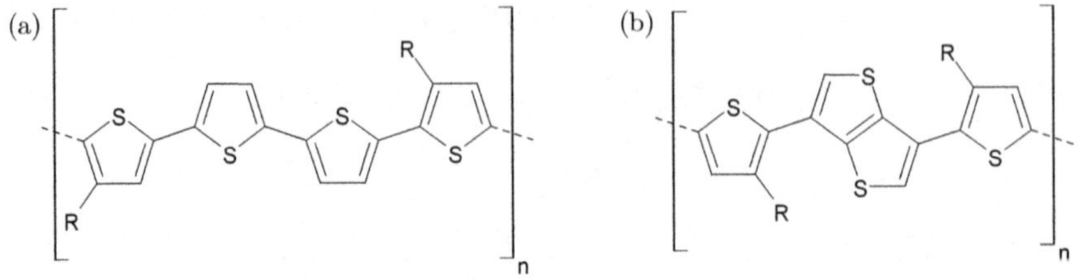

Fig. 3.4. Two advanced polythiophene derivatives with liquid crystal properties. These materials are solution processable and can be processed into relatively large crystal domains through controlled thermal cycling. (a) PQT, where R is an alkane (in PQT-12, for example, R is $(CH_{11})CH_3$) [21] (b) poly(2,5-bis(3-alkylthiophen-2-yl) thieno[3,2-b]thiophene), where R is also a variable length alkane chain 10-14 C long ([22])

beyond that achievable with regioregular P3HT or PQT-family materials and exhibits excellent performance [22]. Other combinations have been and continue to be developed with better performance and processing characteristics.

Polyfluorenes

Another popular category of polymeric semiconductor materials is the polyfluorenes, which have the general structure shown in Fig. 3.5 (a) [23]. The variant F8T2, which is shown in Fig. 3.5 (b) has attracted the greatest attention because of its easily accessible nematic transition at $\tilde{2}80^{\circ}$C and high solubility due to its alkyl chains.

Polyfluorenes have also been engineered for a high degree of liquid crystallinity. Through copolymerization with other units (like thiophenes or vinylidines) the lamellar packing, HOMO-LUMO gap, and HOMO/LUMO alignment can be engineered. Like other liquid crystalline polymers the crystallization through controlled heating, substrate templating, and quenching plays an important role in achieving good device performance.

3.2 Small molecule semiconductors

As with polymeric organic semiconductor materials, the variety of available small molecule materials is almost infinite. Many materials are chain or fused units of the fundamental building blocks shown in Fig. 3.1. Small molecules are light and can usually be purified and deposited through thermal evaporation processes. Many small molecule organic semiconductors have also been functionalized to be soluble and can be deposited from solution.

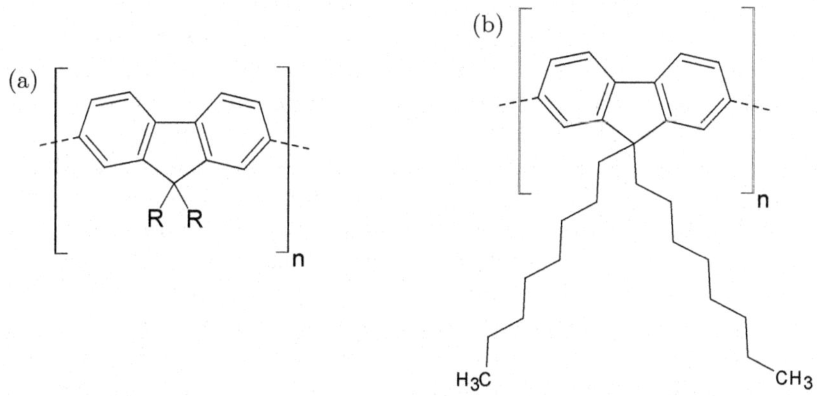

Fig. 3.5. (a) The general structure of polyfluorenes, and (b) the variant F8T2. Polyfluorenes and many of their derivatives are liquid crystals which can be crystallized after deposition to improve the pi-pi overlap and device performance. The alkyl chains tethered to the fluorene group solubilize the material.

Acenes and their derivatives

Fused ring structures form a large variety of organic semiconductor materials. May of these fused ring assemblies are planar and rigid, which leads to superior stacking properties. Many materials will form polycrystalline films when deposited at or near room temperature. This better order, which can have better pi-pi overlap between neighboring molecules, improves transport and overall OFET performance.

The acenes are a series of fused benzene rings and are shown in Fig. 3.6.

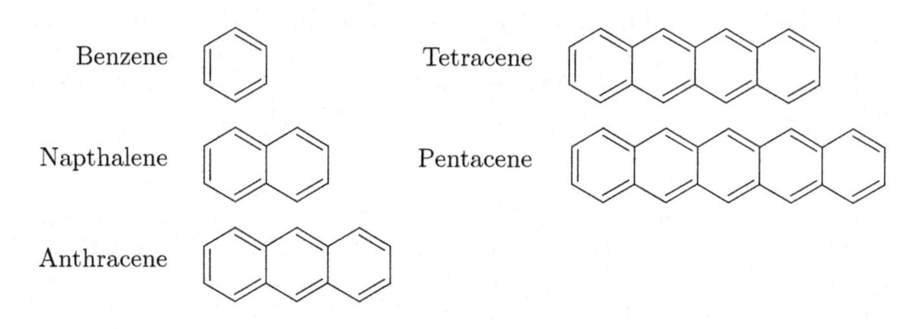

Fig. 3.6. The first five acenes.

Tetracene and pentacene are particularly interesting small molecules for OFETs. Both strongly organize to form polycrystalline films with good transport properties on insulating substrates and can also form large single crystals under appropriate conditions. Pentacene is by far the most popular organic semiconductor for OFET fabrication.

Despite this popularity, pentacene is not ideal. Pentacene is not soluble to an appreciable degree and can only be deposited using vacuum processes. Pentacene is relatively easily oxidized, especially at the 6,13 positions, which disrupts transport and crystallization in devices. Finally, pentacene can condense into two crystal phases which are closely related, but not perfectly matched. This often leads to polymorphic crystal growth, which can lead to mismatched grains and decreased OFET performance. Several approaches have been pursued to rectify these problems.

One strategy to overcome all three objections simultaneously is to synthetically attach bulky groups to the 6,13 positions of pentacene. When the groups are properly selected this passivates the most reactive sites of the molecule, constrains the crystallization into a single highly favorable phase, and imparts solubility [24]. While soluble, TIPS pentacene is also light enough to be purified and deposited using vacuum sublimation. The structure of TIPS pentacene is shown in Fig. 3.7 (a).

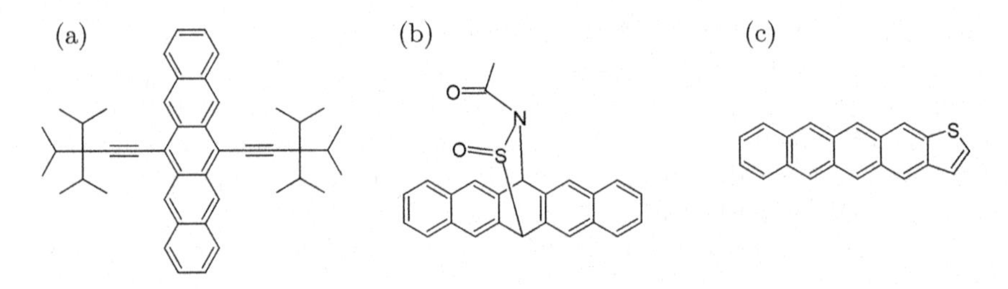

Fig. 3.7. The structure of several functionalized pentacenes. (a) TIPS-pentacene, one of a family bulky group functionalized pentacenes developed by Anthony et al. [25]. (b) A soluble conversion material which can be printed and then converted into an insoluble transistor through heat exposure [26]. (c) A thiophene containing pentacene analog which incorporates a thiophene ring for greater stability [27]

Several other pentacene derivatives have been developed to add functionality to the basic pentacene unit. Afzali et al. has developed a family of soluble pentacene precursors [26]. These materials can be converted into well packed pentacene films on exposure to heat. The conversion process can also be engineered to be photopatternable. Tang et al. [27] has shown several pentacene and tetracene analogs that increase stability by incorporating thiophene groups at the ends. This decreases the conjugation length which separates the HOMO and LUMO, improving the oxidative stability of the molecules. Since the shape is similar, these analogs retain many of the same favorable crystal-forming kinetics that pentacene and tetracene enjoy.

A number of other related fused ring semiconductors have also attracted significant interest, especially perylene and rubrene. These materials form high quality single crystals and single crystal OFET devices.

Oligothiophenes and oligofluorenes

Oligomeric versions of semiconducting polymers are often good transistor forming materials. Oligothiophenes (such as hexithophene) behave much like the acenes; they form polycrystalline thin films with some pi-orbital overlap parallel to the substrate. Several end group substitutions have been developed to improve the film ordering and stacking, two representative examples are shown in Fig. 3.8.

Oligofluorenes have also been synthesized which retain many of the liquid crystalline properties of their polymer counterparts while allowing vacuum purification and deposition [28].

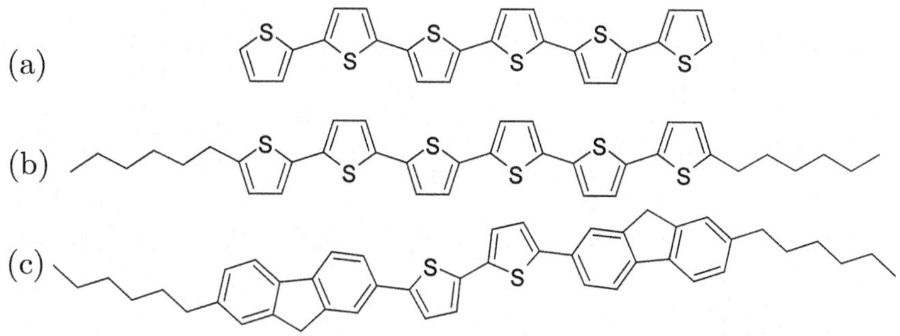

(a)

(b)

(c)

Fig. 3.8. (a) α-hexithiophene, an oligomeric form of polythiophene. This material is typically vacuum deposited. (b) Adds two alkane groups at the ends, which improves the material crystallinity [29]. Despite the methyl groups the solubility in common solvents is low. (c) An oligofluorene with strong self-organizing properties when deposited on a heated substrate [28]

Electron transporting materials

All of the materials presented so far are hole transporting (i.e. p-type); a conjugated carbon backbone carries the charge and is ringed primarily by hydrogens. These hydrogen atoms are less electronegative than the core carbon atoms and donate a partial electron charge to the molecular core. This makes the backbone electron rich and shifts the energy levels so as to favor the loss of an electron over its gain (and, by extension, the formation of a hole channel over an electron channel). Put another way, the negatively charged form of the material is energetically unfavored.

To make transistors that carry electrons easily the conjugated backbone needs to be electron deficient so that the LUMO level falls lower and gaining an electron becomes possible. A number of organic semiconductors have been designed with this functionality in mind, four representative examples are shown in Fig. 3.9, along with C60, which is natively electron transporting.

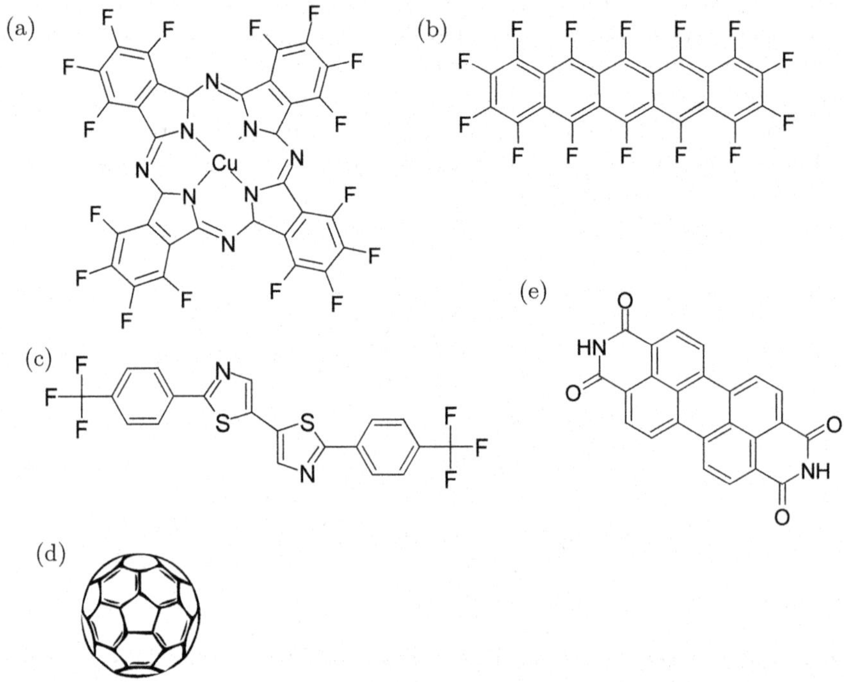

Fig. 3.9. Four representative n-type OFET materials. (a), (b), and (c) achieve electron deficient carbon backbones through fluorination, although other electronegative groups may also be used. (a) Hexadecafluoro copper pthalocyanine [30], (b) a perfluorinated pentacene [31], and (c) a fluorinated thiazole-based oligomer [32]. (d) C60 is also an electron transporting material which can form OFETs [33]. (e) is a perlyene derivative, perylene 3,4,9,10 tetracarboxydiimide, which has found wide application as an n-type material in organic solar cells. A large number of n-type perylene derivatives have been developed.

N-type organic semiconductor materials are generally not air stable, for reasons discussed in the next section, but several flourinated compounds (such as F16CuPC) are natively relativle stable, and several newly developed n-type materials have been engineered to be air stable [34].

Extrinsic doping with electropositive materials (e.g. alkali metals) and contact doping with low work function materials has also been proposed to create n-type devices. While this strategy works on a number of materials (including pentacene), direct LUMO injection doping has not progressed to a point where it is straightforward to use. The dopants used have a tendency to migrate under bias, and for the same reasons that make these materials easily donate electrons to the semiconductor, also tend to be highly reactive with air due to their low ionization energies. Further development of dopant materials and/or encapsulation strategies are needed for this approach to enter mainstream use.

3.2.1 Air stability

A major complication of both n- and p-type organic semiconductors is their poor stability against oxidation and reduction reactions.

De Leeuw enumerates the oxidation and reduction reactions that can occur against an organic semiconductor in [35], and compares the reaction potentials of likely interactions (i.e. those involving water and oxygen) with the HOMO and LUMO energies of several organic semiconductor materials. In the presence of water and oxygen, there are H+ ions, OH-ions, and O_2 molecules available for oxidation or reduction. By looking at the reduction potentials of reactions involving just water ions or water together with oxygen, he demonstrates that the most likely reactions for a non-air stable n-type material is reduction of H+ and oxygen into H_2O. This leads to conversion of a negatively charged semiconductor unit into a neutral unit and charged hydrogen and OH radicals which disrupt transport in the device. Charged p-type organic semiconductors are most easily attacked in a reaction which reduces water.

By engineering the electron energy levels of organic semiconductors it is possible to set the HOMO and LUMO levels so that the equilibrium potential at no bias sits in a region which is thermodynamically stable against these redox reactions. For p-type materials, experiments have shown that the HOMO needs to lie more than 5.2V below the vacuum level to have this stability [36]–a goal which is achievable with a number of materials. To create an n-type material which is stable against oxygen is more challenging. N-typ stability requires an oxidation potential greater than 0.57V vs SCE which is synthetically difficult but possible (see, for example, [37]).

It is also possible that steric interference with oxygen and/or water can also stabilize OFET materials. A practical consequence is that for devices whose energy levels lie outside the stability range it is possible to have devices which are unstable against oxygen alone, or devices which are unstable against oxygen in the presence of water. This can inform the type of passivation which may be used to protect the device against environmental degradation.

Because an applied bias can change the local potentials, some marginally stable materials may require dehydration and passivation to achieve greater stability under bias.

3.2.2 Organic conductors

Many organic semiconducting polymers can be doped to form highly conductive materials with a variety of uses. These materials are often solution processed, which allows printing or other solution coating to be performed. When heavily doped, these materials can have conductivities approaching those of highly conductive metals.

Many unsubstituted polymeric organic semiconductors are not easily dissolved in solvents. After mechanical processing, doping with acids, and/or processing with surfactants many of these materials can be formed into conductive dispersions and applied from solution to substrates.

These materials can be used to fabricate contacts to OFETs. In many cases the energy band alignment and material growth on organic conductor source and drain electrodes is better than that observed on metal contacts. These materials are typically insoluble but can be produced as metastable dispersions through high energy processing techniques and appropriate selection of dopants and surfactants. These dispersions can then be printed or blanket coated and subtractively patterned or locally de-doped to define circuit elements.

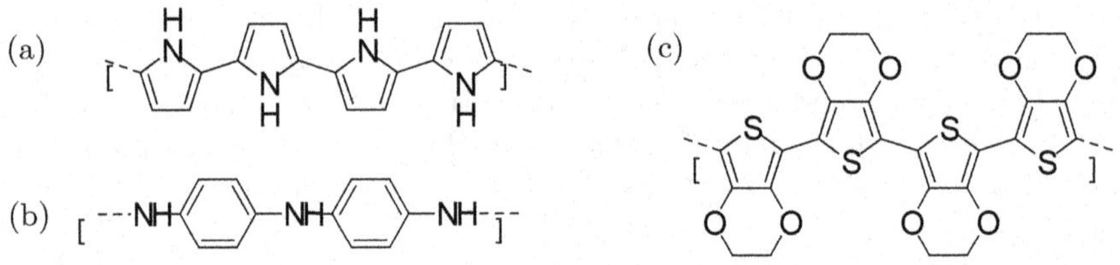

(a) (b) (c)

Fig. 3.10. Some of the more commonly encountered organic conductor materials (a) polypyrrole, (b) polyaniline, and (c) poly(3,4-ethylenedioxythiophene) (PEDOT). When combined with water soluble organic acids (e.g. sulfonic acids like benzosulfonic acid) many of these polymers can form doped complexes which are highly conductive and can be dispersed into suspension. Substituted versions of these polymers which are self-doped have also been developed.

3.3 Conclusions

Organic synthetic chemistry has yielded a tremendous variety of materials suitable for use in OFETs. Materials with improved processability, performance, stability, and throughput have been developed through the application of rational chemical design strategies and advanced synthetic techniques. It is certain that new materials with superior properties and devices which can exploit them will continue to be developed in academic and industrial labs around the world.

4

Basic OFET fabrication

4.1 Introduction

Organic semiconductors, because of their van der Waals bonded structure, have relatively weak interactions between molecules. This makes it possible to transport and manipulate organic semiconductor materials without the need to break or re-form covalent bonds. It is therefore possible to manipulate molecules of these materials using a modest energetic input, since the activation energy required to make and break van der Waals bonds is less than that of most covalent bonds. It is also possible to avoid forming a large density of electrically active broken covalent bonds, since whole molecules can be moved and reconstituted into solid films while retaining their semiconducting properties.

This weak intermolecular interaction allows processing using a number of techniques unavailable to covalently bonded thin film materials (such as thermal vacuum deposition and solution processing), but also means that the structure and properties of organic semiconductor layers can easily be disrupted. This chapter will discuss some of the processing strategies which have been developed for fabricating OFETs in light of the energetics and chemistry of the more popular material classes.

4.2 Basic OFET structure and operation

OFETs operate through the creation and elimination of a sheet of charge carriers at the gate dielectric/body interface (this is further discussed in chapter 6). The basic operation is summarized in Fig. 4.1.

In a traditional OFET, the charge carriers travel in the channel parallel to the surface of the gate dielectric. The source and drain electrodes provide access to the channel and are engineered to inject charge well under all bias conditions. It is the gap between the source and drain–the channel–whose conductance is switched, adjusting the conductance of the whole device. The

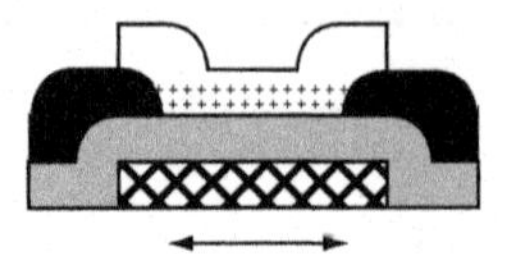

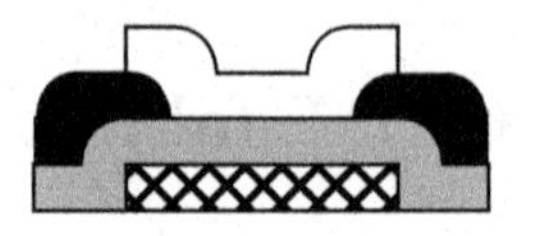

Fig. 4.1. A traditional four layer OFET design. When the device is biased so that a sheet of charge is formed between the source and drain, current can flow through the channel when a bias is applied between the source and drain, the details of which depend on the magnitude of the bias. When the device is biased in an operating region which eliminates the sheet charge in the channel, the device is insulating and little current flows, independent of the applied source/drain bias.

gate is separated from the channel by the gate dielectric forming a capacitor to the channel charge sheet. It is the gate dielectric that allows the creation of a field across the semiconductor and the resultant accumulation and depletion of carriers without the need for a DC current. The unit operations and process flows described in this chapter are the building blocks for many processes used to fabricate OFETs. A number of designs which do not follow this general structure have been developed, several are discussed further in Section 5.6.

4.3 Unit operations

4.3.1 Thermal evaporation

Technique

Thermal evaporation in vacuum allows for deposition and purification of small molecule organic semiconductors. The material is placed into a vacuum and heated. Once the vapor pressure of the heated material exceeds the background pressure of the material in the chamber, the material evaporates and condenses on cooler surfaces that it lands on. High molecular weight organic semiconductors cannot be deposited this way; they are too heavy to evaporate and decompose instead.

Thermal evaporation of metals and semiconductor materials is typically performed under high or ultra high vacuum conditions. This is done for three main reasons. First, the evacuation of the chamber reduces the partial pressure of oxygen or other gases which may react with, become embedded in, or otherwise contaminate the sample or heating apparatus. Second, the vapor pressure of most solids is low, and evaporating in a vacuum allows for these low vapor pressure materials to evaporate. Finally, the mean free path of the evaporated material is extended to larger than the source-substrate dimension, allowing the evaporant to travel to the substrate without deflection.

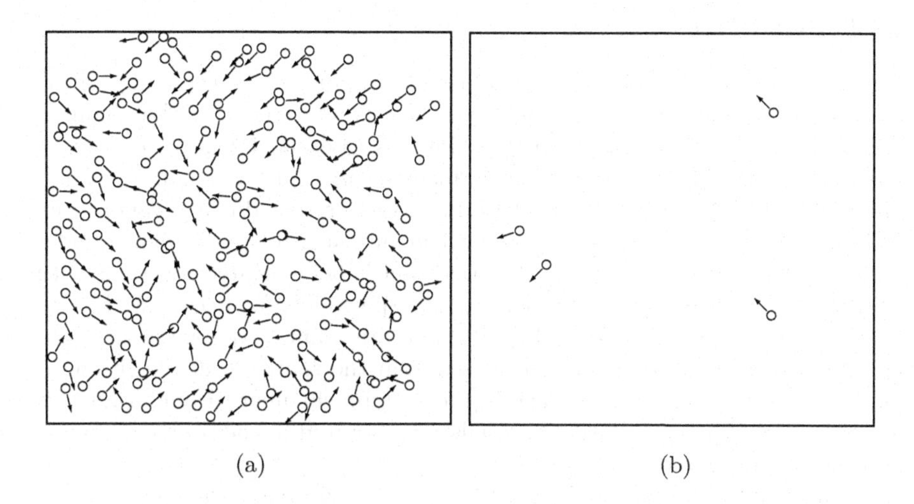

Fig. 4.2. A schematic showing the presence of gas at (a) higher and (b) lower pressures. The pressure is a function of the number of collisions with the container by the gas molecules and the average force that each collision carries. The molecules also collide with each other, and the mean free path of molecules in the gas is also a function of the gas molecule density (and therefore pressure).

Gases can be seen as collections of particles that collide elastically with each other and have a total kinetic energy that depends on the temperature of the gas. The mean free path for a gas can be shown to be proportional to the pressure and is given by:

$$\lambda = \frac{kT}{4\pi\sqrt{2}r^2p} \tag{4.1}$$

where r is the effective molecular radius and p is the pressure in pascal. For air (and pressure in torr) this can be approximated as:

$$\lambda = \frac{0.005cm}{p} \tag{4.2}$$

In air and at room temperature, a typical mean free path is $65nm$. Evaporated materials at atmospheric pressure, therefore, scatter many times before traveling any appreciable distance and the directionality of their departure is lost. Typical pressures for vacuum thermal evaporation are 10^{-6}-10^{-8} torr. A typical mean free path is on the order of meters or longer–at 10^{-8} torr, the mean free path is approximately 5km. This allows for evaporant materials to travel in a line-of-sight pattern from the source to the substrate.

Background gases

An important consideration during thermal evaporation is the composition of the background gas during the deposition. Management of the chamber pressure during pre-heating and deposition is essential for good results. It has been shown that organic semiconductors can be intentionally [38] and unintentionally [39] doped by background gases present during evaporation. Kelley, et al. also show a clear improvement in transistor performance under higher chamber vacuum [40]. As the evaporant is being heated reactions with water, oxygen, hydrogen, and other reactive species can be accelerated.

Once heating of the source begins there are additional sources of background gas including desorption of gases from the source, crucible (if one is used), filaments, and other surfaces in the vacuum chamber through radiative or conductive heating; volatile materials incorporated in the source material; and thermal decomposition products.

Some evaporants (especially reactive metals) will react with gases which are in the chamber. A clear indication that this is occurring will be a reduction in the chamber pressure while the material is being evaporated. While this can be favorable from the perspective of creating a good vacuum (and there is a vacuum pump which is based on this principle), reactive metals deposited in a high pressure ambient can have different properties than expected because of their reaction with background gases.

Resistively heated foil boats

The simplest heating system runs large currents through a metal foil boat which is stamped, rolled and pierced, or folded to contain the evaporant source material. Sheets of refractory metals such as molybdenum, tungsten, and tantalum are commonly used, but other materials such as aluminum foil may also be employed, especially at lower temperatures. The boat heats up through resistive loss and heat is lost through radiation, the heat of fusion and vaporization of the source, and conduction into the electrodes. The electrodes are necessarily high conductivity metals (typically copper blocks) and conduct heat well. The equilibrium temperature is achieved when the energetic input and loss are balanced.

Many metals can be evaporated using this arrangement and some more sophisticated boat designs have been developed to allow a wider range of evaporation conditions. It is possible to coat the boat with dewetting materials (e.g. Al_2O_3) to confine the molten source material, neck the boat down to

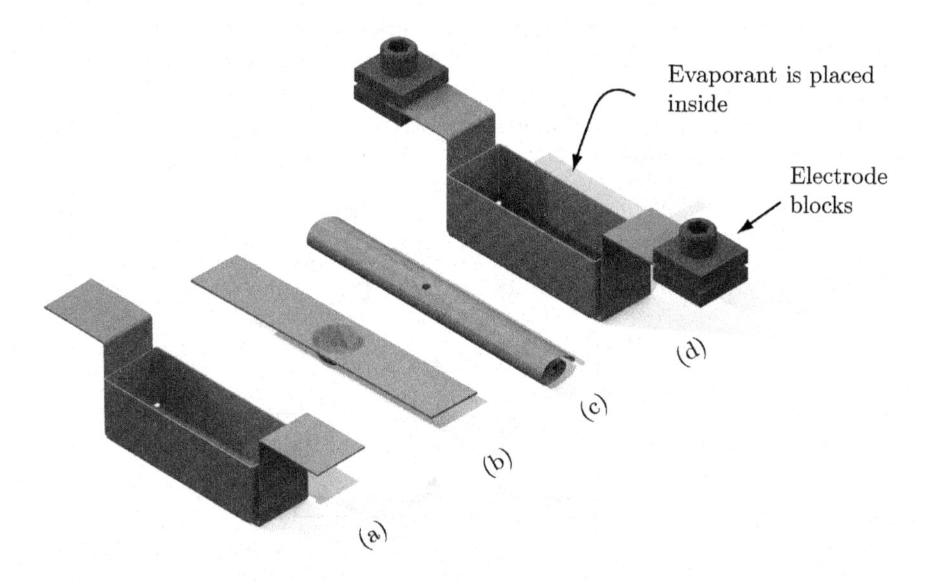

Fig. 4.3. Schematic showing several styles of resistively heated boats. (a) is a folded boat, (b) is a stamped boat, and (c) is a rolled foil packet with a small hole punched in the center. (d) shows the folded boat clamped to the copper electrodes; the electrodes also heatsink the boat.

decrease the thermal conductivity, and to include some radiation shielding in the boat structure.

Materials which melt during the heating process pose a challenge for thermal evaporation. The source material is loaded in a solid form which has poor thermal and electrical contact with the boat. After chamber evacuation and heating, once a portion of the evaporant melts, significantly better thermal and, in the case of metals, electrical contact is achieved. This usually causes the rest of the charge to melt almost instantly and, in the case of metals, causes a drop in the boat resistance which can lead to a current surge if only voltage is regulated. This requires some care in evaporation. The use of ceramic coated boats, which do not make electrical contact with the evaporant, can mitigate this problem.

For many materials (including organic sources, which typically sublime at relatively low temperatures), the evaporation source may be used an unlimited number of times. Upon cooling, materials which melt may experience a thermal expansion mismatch which can cause flaking or delamination from the boat. When using evaporants which melt refilling pre-conditioned sources with molten evaporants while there is a pool of solid material in contact with the boat can lead to greater process uniformity.

Materials which sublime can experience a 'popcorn' effect in which the part of the source granules in contact with the substrate vaporizes first and the expanding gas causes ejection of the material. In mild cases this causes a loss of material, in extreme cases it can deposit particles on the substrate and cause premature source depletion. This problem can be overcome by using a baffled source in which evaporants do not travel via a line of sight to the substrate.

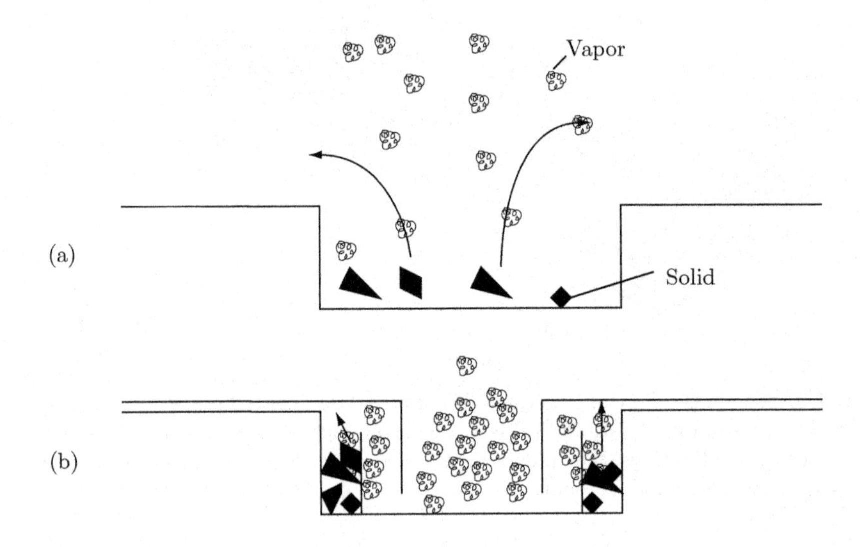

Fig. 4.4. A schematic cross-section of two evaporation boats showing the popcorn effect. In (a), the boat is open, and the evaporant material sublimes first at the hottest point of the assembly, which is the bottom of the material granules. This creates a vapor jet which can eject solid particles of evaporant from the boat. In a baffled arrangement (b), the source is constructed so as to prevent line of sight contact between the source and the substrate. Solid particles which are propelled are trapped, whereas material vapor can escape and deposit.

Aluminum is a particularly difficult material to evaporate using foil boats. Molten aluminum alloys with the materials commonly used to make foil boats, leading to embrittlement and breakage. Many solutions have been proposed, none are totally satisfactory. One popular approach is to stack two thin foil boats so that the top boat is sacrificial and the bottom boat contains the evaporant. It is also possible to use a thick boat and replace it every few uses before failure.

Braided wires can also be used in the same configuration as stamped boats; this is most effective for metals which melt, although large chips of metals which sublime also can evaporate well from coiled braids. Molten metals are drawn into the wire via capillary action and then evaporate. Because braided wires have a higher electrical and thermal resistance than wide foil boats, and therefore can achieve a higher temperature, evaporation of metals which may be difficult on some equipment using boats (e.g. Ni) is possible using braids.

Crucible evaporation

Material can also be evaporated from crucibles, which is in many ways the same process as that used for boat evaporation. The major difference is that the crucible (which can be ceramic, quartz, graphite, glass, or metal) is indirectly heated by a resistive heater. This has some advantages, especially for the deposition of organic semiconductors, but can also be used to deposit many metals and some insulator and optical materials.

The simplest crucible heaters are made from braided tungsten filaments fashioned to hold the crucible. A related style uses a formed foil boat to hold the crucible.

An advanced style of crucible evaporator is a Knudsen cell (K-cell). By measuring the flow rate of rarefied gases Knudsen showed that the reflection of gas molecules from the chamber walls does not retain any directionality [41]. A consequence of this is that when the mean free path of gas molecules exceeds the exit aperture of a crucible by a factor of 10 or so and there is no line of sight path from the evaporant to the orfice, reflection of the evaporant from the volume sidewalls occurs randomly and the exit orifice appears as a Lambertial virtual source of evaporant [42]. This is the gas kinetic equivalent to how an integrating sphere creates a virtual Lambertian source of light by redirecting the incoming light source off of a series of specular scattering surfaces.

A ideal Knudsen cell holds the source in an isothermal box with a relatively small exit orfice. This can be implemented by placing the crucible and filament inside a radiation shield without making contact between the three elements. The filament heats the crucible radiatively (there is no gas or direct mechanical connection to conduct heat), and this multiple reflection and conduction inside the crucible homogenizes the temperature along the crucible length. This configuration also increases the temperature that can be achieved for a given power input, since less heat is lost to the environment via radiation. The maximum temperature which can be achieved by a K-cell is limited by the metal elements used to fabricate the cell and the thermocouples, and is typically 1200-1400°C.

In practice a perfectly uniform cell temperature profile is not desirable. The temperature gradient can be manipulated inside the cell by placing more windings at the lip or creating multiple heating zones. It is generally desirable to heat the exit lip more than the rest of the crucible to prevent clogging and to compensate for heat which is preferentially radiatively lost at the cell end [43]. Monitoring and controlling the source temperature for feedback is more straightforward in a shielded crucible environment since the heating element itself is not being measured (e.g. while it is possible to weld a thermocouple to a foil boat this has many sources of error). The more uniform, omnidirectional, and controlled heating achievable with a crucible also minimizes particle ejection through the popcorn effect since the source material is radiatively heated from all sides.

Several specialized heaters designed for organic semiconductors have been developed and sources designed for molecular beam epitaxy (MBE) can also be used. The directionality desired can be engineered by adjusting the crucible aspect ratio. Sources with a small effusion orfice at the lip or large depth operate in a more ideal Knudsen regime and coat uniformly over a spherical profile; these are suitable for substrates held in a planetary configuration. Other sources are designed with larger exit orfices relative to the crucible depth and achieve a more directional deposition (since there is a greater line-of-sight contribution) which can increase deposition efficiency and uniformity on planar and smaller samples.

A typical crucible heater design is schematically shown in Fig. 4.5.

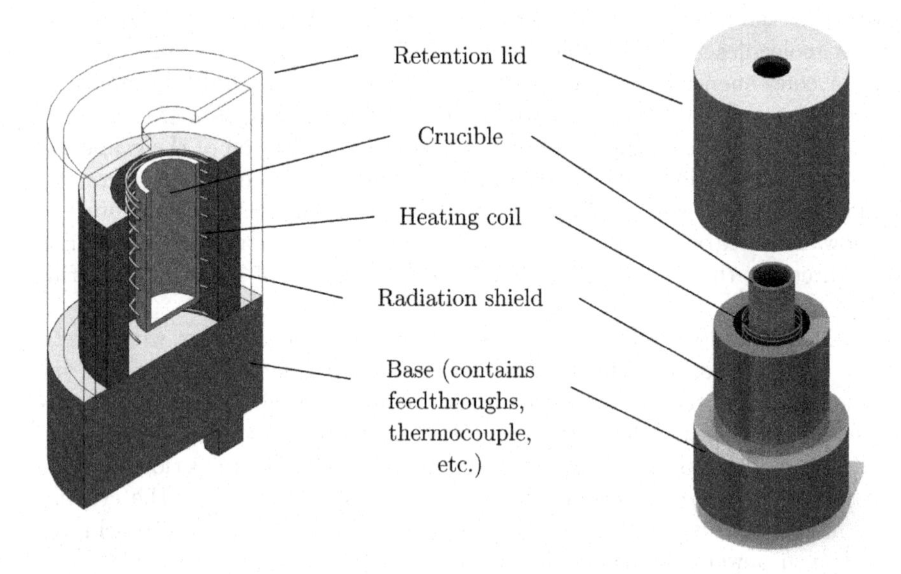

Fig. 4.5. An exploded and cross-sectional schematic view of a heater designed or evaporation from a crucible. The coils are more tightly packed towards the top of the heater to create a heated lip to prevent material accumulation on the top of the unit.

Substrate temperature

As will be discussed in Section 4.4.2, the crystal growth process for many organic semiconductors requires control of the substrate temperature and deposition rate to manage the crystal structure and incorporation of impurities.

The infra-red radiative flux from the evaporation source and heater can cause a significant increase in the substrate temperature. This problem is worse with large area unshielded evaporation sources (i.e. foil boats) which emit a significant amount of IR radiation from a large area. All resistive evaporation will emit some IR radiation and the substrate temperature will climb during deposition. If substrate temperature control is critical, a shielded crucible source or a substrate stage which can both heat and cool (either actively or through conduction to a cooling block) may be required.

Electron beam evaporation

Electron beam (e-beam) evaporation is a technique suitable for evaporating high temperature evaporation materials such as metals, glasses, and ceramics. It is not used for evaporation of organic semiconductors, which have substantially lower evaporation temperatures.

The source material is contained in a crucible held under vacuum. A heated filament emits high voltage electrons thermionically, and a focused beam is launched into the vacuum and steered into the material charge via magnetic optical elements. The electrons impact the substrate, lose their kinetic energy, and cause substantial heating. The energy density of the beam can be controlled through management of the beam kinetic energy and current, and also by modulating the point of impact through magnetic beam steering to cover a larger or smaller area of the evaporant.

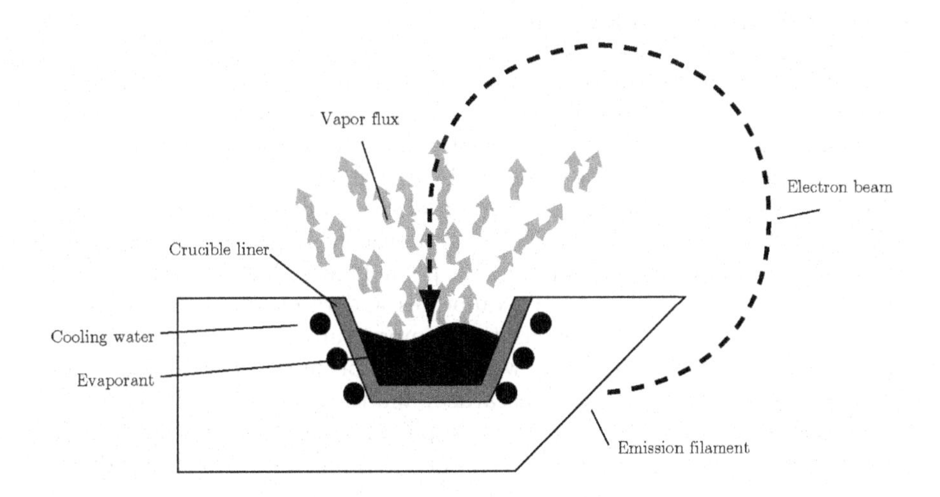

Fig. 4.6. Schematic cross-section of an electron-beam gun for e-beam evaporation.

Electron beam deposition can reach extremely high temperatures, well over 3500°C, which allows evaporation of materials with very high evaporation points, including refractory ceramics. By using a temperature gradient it is possible to evaporate materials that would otherwise melt structures designed to hold the material. The pocket cooling establishes a temperature gradient which holds the molten material in a crucible of the solid form of the same material. Lower melting point materials (such as gold and silver) can be placed in crucibles which the molten material does not wet and the entire charge can be melted. Source material and crucible manufacturers offer a variety of guides recommending crucibles for materials. Because the deposition occurs in high vacuum and can be engineered to avoid any melt contact with a crucible, the purity of the evaporated film can be high.

There are several potential disadvantages to electron beam deposition:

- Samples can experience a significant flux of reflected electrons, which can charge or otherwise disrupt some materials such as e-beam resists
- Because a heated filament is used to create the electron beam, the background pressure must be held low to avoid rapid oxidation of the filament
- The hot substrate can emit a substantial IR and UV radiation through blackbody emission, as well as some Bremsstrahlung x-ray radiation
- The technique is restricted to materials which have a high decomposition temperature and can tolerate electron bombardment; some complex materials (including many carbides and oxides) can survive the process with their stoichiometry relatively intact, but others decompose.

Thermal gradient evaporation

The purification of organic semiconductors [44] and the fabrication of large single crystals [45] is typically performed using a controlled atmosphere or evacuated tube together with an external resistively heated furnace. This approach is a descendant of that proposed by McGhie et al. [46].

The basic principle for temperature gradient sublimation is summarized in Fig. 4.7. The source material, held in an evacuated tube or an inert gas, is placed in a temperature gradient at a point higher than the evaporation temperature. The material slowly evaporates and recondenses at the point at which the tube is at the condensation temperature. The higher and lower evaporation temperature fraction (which consist of impurities and decomposition products) is separated and discarded, and the process can be repeated. When performed with a background atmosphere, inert or reactive flowing gas through the tube can effect an additional material flux and speed the process, and can also manipulate the vapor saturation of the gas and engineer crystal growth.

The material which condenses has a high kinetic energy and the ability to exclude impurities through successive recrystallization. This allows material of high purity to be achieved. Large single crystals of a number of materials (including rubrene, tetracene, and pentacene) have also been demonstrated with exceptional physical and electrical properties (see, for example, [47]). A material flux driven in a related technique has also been used to deposit and pattern organic semiconductors in a highly controlled fashion [48].

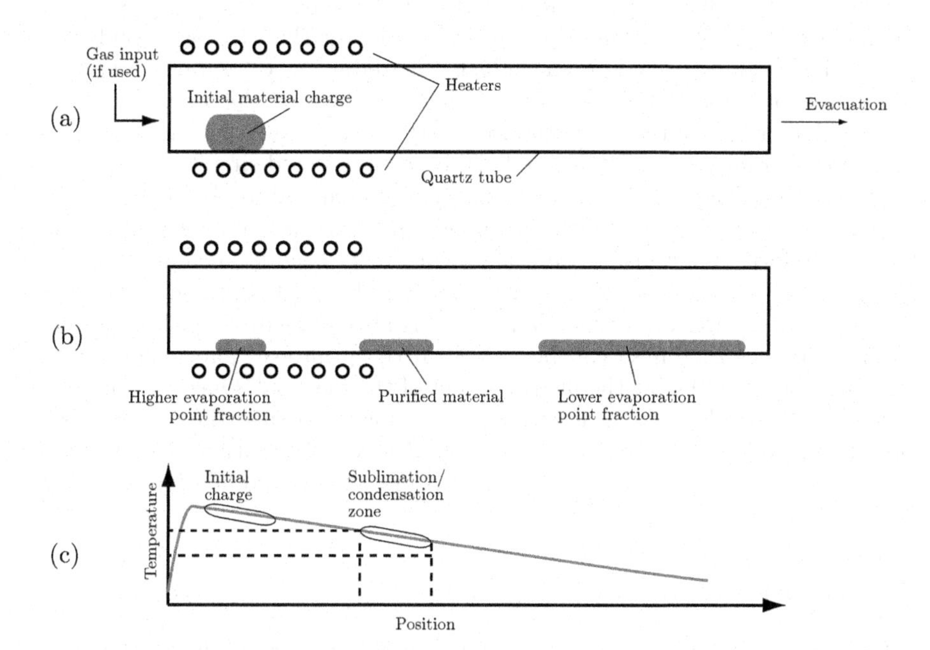

Fig. 4.7. A schematic showing the fundamentals of temperature gradient sublimation for organic semiconductors. (a) shows the tube setup before heating; the material charge is placed in a region of negative thermal gradient slope to prevent retrograde motion. The tube is evacuated or filled with the desired atmosphere. If vacuum is used the tube may be evacuated from either end, otherwise a convective flow in the direction of the decreasing gradient can be established to accelerate the process. The temperature gradient shown in (c) is then applied by heating one end and cooling the far end actively or allowing it to cool through atmospheric convection. (b) shows the final product when the process is complete; the purified fraction will condense at the sublimation temperature, and higher and lower vaporization point fractions are spatially separated and the purified material can be recovered.

The ability to purify and crystallize material using this class of techniques and readily available equipment is an advantage of small molecule organic semiconductors over polymeric materials.

4.3.2 Liquid deposition

Liquid deposition processes are an important part of most OFET fabrication processes, either to deposit the active layers, or to manipulate layers deposited through other means. Because the individual molecules in organic solids are weakly bonded, these bonds can be dissociated through exposure to solvents which have a greater affinity for the semiconductor than the semiconductor has to itself. Many organic semiconductor materials have been engineered to be soluble or dispersible in solution, taking advantage of this property.

Insulators and conductors can also be deposited using solution deposition to form other layers in OFETs or OFET-based integrated circuits. Dispersible metallic and ceramic nanoparticles can be deposited and used directly or sintered to form high quality films. Precursor materials can also be deposited and reacted to form thin films. A range of precursors for conducting and insulating polymers, ceramics, and metals exist which can be solution patterned.

Solid waxes are a material class which is also compatible with liquid deposition processes. These can be heated in the printhead and deposited on the substrate held below the melting point of the material. Once the material prints it freezes–by taking advantage of the phase change there is no need for drying and no solvent evolution or underlayer dissolution. These liquid deposited waxes can be used as etch resists and affinity modulation layers for patterning.

Blanket coating

Many strategies have been developed for blanket solution coating for the semiconductor and printing industries. A large number have been applied to the deposition of organic semiconductors, organic insulators, and solution processable metallic layers for use in OFETs.

The more common deposition strategies include:

- Spin-coating
- Drop-casting
- Dip coating
- Spray coating
- Blade coating techniques (e.g. doctor blading, etc.)
- Roll coating techniques (e.g. graveure and its variants)

These have been reviewed in the literature and their relative merits compared [49].

When used to deposit blanket coatings, these films usually need to be patterned using another technique, although some can be modified to accomodate patterning during the deposition process.

Printing

Printing comprises a family of techniques that simultaneously deposit and pattern a target material.

Printing techniques include:

- Ejected drop printing (e.g. through piezo or thermal inkjet)
- Pneumatic drop or stream dispensing
- Contact stamp printing
- Indirect and offset printing methods
- Capillary stylus dispensing

There are many potential advantages to printing one or more of the layers in a device including:

- Elimination of a separate patterning step
- More efficient use of material
- Device patterning without fixed tooling (since the printing pattern is usually software controlled)
- Potentially high throughput through use of parallel deposition processes

As with blanket deposition, printing has been used to deposit a full range of materials including conductors, insulators, semiconductors, masking materials, and surface energy modulation materials to create OFETs and integrated circuits with OFETs.

Piezo inkjet printing has dominated OFET fabrication using printing techniques due to its excellent chemical compatibility and the availability of sophisticated printheads to the development community.

4.3.3 Polymer CVD

Chemical vapor deposition (CVD) of polymers allows the formation of polymer thin films directly on substrates from a precursor vapor stream of monomers or oligomers. Two of the more common are briefly described.

Hot wire and plasma

In traditional polymer CVD, the precursor stream flowing with a carrier gas is fed over a heated tungsten wire (hot wire) or through a microwave cavity (plasma) in a reduced atmosphere chamber. The activated monomers then condense into a polymer film on nearby surfaces. This technique can be used to create high quality insulators [50] as well as doped organic conducting polymer materials [51] by proper selection of the source materials and the use of suitable catalysts.

Parylene

Parylene is a unique CVD deposited insulating polymer, first discovered in 1947 [52]. The material is supplied as a metastable dimer first described by [53]. The dimer is sublimed under vacuum and flows into a hot wall furnace where it is cracked into a pair of direactive monomer units. These then flow by diffusion into the deposition chamber, which is near or below room temperature, and condense into a high molecular weight polymer. The polymer is fully reacted without the need for any post-processing.

Because the deposition occurs at room temperature the thermal mismatch, strain, and other complications associated with other CVD techniques are not observed.

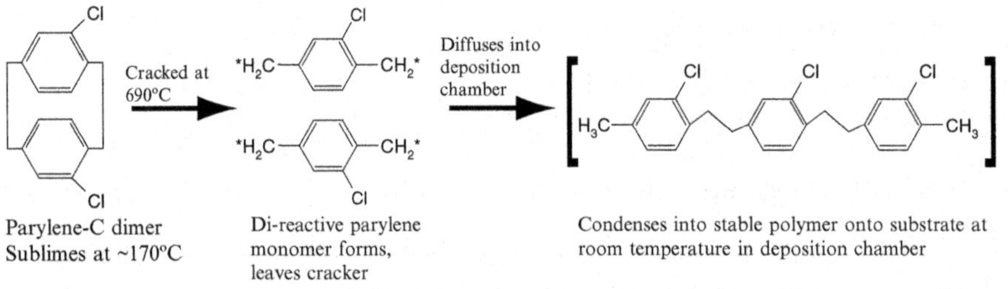

Cracked at 690°C

Diffuses into deposition chamber

Parylene-C dimer Sublimes at ~170°C

Di-reactive parylene monomer forms, leaves cracker

Condenses into stable polymer onto substrate at room temperature in deposition chamber

Fig. 4.8. A summary of the three steps of the parylene deposition process. Parylene-C is shown. First, the stable dimer is vaporized at approximately 170°C, and diffuses into the cracker, which is held around 700°C. The cracker breaks the dimer into two reactive monomer units which diffuse into the deposition chamber and condense into a fully reacted polymer at room temperature. Unreacted material is captured in a cold trap to protect the vacuum pump, and the deposition process occurs at approximately 10-50mtorr for parylene-C.

Several types of parylene dimer are commercially available (and many more have been synthesized and reported in the scientific and patent literature), the most popular are C, D, and N. Parylene-C is often used as an encapsulant for organic semiconductors, and is also sometimes used as a gate dielectric.

4.3.4 Other applicable PVD and CVD processes

Many of the other techniques applied to other semiconductor and printing systems can also be applied to OFET processing; an exhaustive description could fill several books. Several processes are worth mentioning, however.

Sputtering, which is the microscopic ablation of a target with an ionized gas, can produce high quality metal and insulator films at high deposition rates. Many interesting materials including high gate dielectric constant materials and transparent conductors can be deposited through sputtering, and virtually all metals (including some which are challenging to evaporate) are easily deposited using the technique. A caveat to using sputtering is that the ionized gas is produced using a plasma, and this plasma can affect the properties of the gate dielectric during the sputtering of the source and drain electrode layer. Oxygen species, intentional or unintentional, in the plasma can cause the formation of potentially undesirable dopant states when sputtering onto a polymeric gate dielectric [54].

Traditional plasma enhanced CVD (PECVD) can be used to deposit insulating films for OFET gate dielectrics. PECVD has been extensively explored by a number of groups and can produce high quality insulators. In applications where the substrate can tolerate the thermal budget required for the deposited film this is a viable approach. High temperature low-pressure CVD (LPCVD) techniques are generally not applied to OFET processes because of their need for high temperature compatible substrates.

Atomic layer deposition (ALD) is a self-limiting process for creating high quality dielectric films. With appropriate process engineering, it is possible to achieve high gate capacitance with low leakage in an additive process. A number of materials, including, Al_2O_3 and HFO_2 have been engineered for deposition with ALD and all indications are that these form excellent transistors even at low processing temperatures. ALD processes have also been applied to organic device encapsulation.

4.3.5 Subtractive patterning operations

In order to form OFETs, it is necessary to pattern at least the source and drain layer of the device. To form circuits, typically four layers need to be patterned. Printing techniques pattern additively, i.e. material is only deposited where it is wanted. When a material is deposited using a large area deposition technique, the unwanted material needs to be defined and removed in a subtractive process.

Photolithography

Photolithography is one strategy for layer definition. In photolithography, a radiation sensitive material is exposed to a structured radiation pattern (e.g. UV through a high optical density mask) and developed. This template is then used as a mask or lift-off vehicle for other materials which are not themselves light sensitive. Photolithography can be directly applied to all of the layers of the organic semiconductor stack. Electron beam lithography can also be used for the same purpose.

One significant problem with the use of photolithography is the difficulty associated with removing photoresist after the etching process is complete. Even slight surface residues can affect device performance, and the stripping/cleaning steps after lithography need to be engineered to insure optimal surface cleanliness. Many organic semiconductor materials are also not tolerant to solvent or developer exposure. This limits the applicability of standard resist systems on semiconductor materials without further engineering.

Polymer semiconductor materials which are soluble cannot tolerate exposure to solvents, which can cause them to swell and/or dissolve. Many oligomeric organic semiconductors are also not tolerant to solvent exposure, which complicates photolithography. Pentacene, for example, is not soluble to any significant degree in any solvents. When pentacene OFETs are exposed to most solvents, however, their semiconductor character is destroyed.

The reason for this puzzling finding was discovered by Gundlach, et al. [55]. When pentacene is deposited it does not crystallize in a single crystal phase, it deposits as a mixture of two phases; one of which is stable at room temperature, and the other which is metastable [56]. When exposed to solvents, the activation energy for conversion of the metastable phase to the stable phase is depressed and the material undergoes a phase transformation into the most stable phase. This causes a dimensional change in the film which cracks and destroys its semiconducting properties. This problem of polymorphic deposition is common to many oligomeric semiconductor materials.

The solution to photolithographically patterning the active layer of OFETs is to protect the semiconductor from materials it is not compatible with (i.e. organic solvents and water). One approach is to use water soluble resists which do not induce the phase transformation [57]. Another is to encapsulate the transistors using parylene first, and then proceed with normal photolithography and patterning [58]. A third approach is to use fluorinated or supercritical CO_2-based resist materials which also do not interact with pentacene or other organic semiconductors [59].

Photosensitization

Another approach to patterning is to sensitize the target material to directly be photosensitive and wash off the remaining material. A number of custom compounded and commercially available materials have been applied as photopatternable gate dielectrics. A similar strategy has also been demonstrated for a soluble pentacene precursor, which converts to an insoluble and well ordered form after treatment [60], and conductor traces in an integrated circuit device [61].

Shadow masking

In shadow masking, a stencil intercepts unwanted material and allows the desired material to reach the substrate. The shadow mask can be a free-standing or tensioned sheet like an etched metal foil, or a patterned layer on the substrate which can be easily removed. The technique is relatively general – evaporation processes, some solution processes, and some sputtering processes can be patterned in this way. The features which can be formed through shadow masking are restricted in their geometry because the mask has to be self-supporting, and there are limits to the practically achievable feature sizes especially as masks grow larger.

Laser ablation

Laser ablation can be used to remove material in a direct write process. It can be applied to metals, semiconductors, and insulating materials with relative ease. Because there are no intermediate processing steps it is a direct and straightforward way to remove unwanted material under computer control. The processing speed and cleanliness of modern laser ablation processes is highly refined and laser ablation is used in a number of large scale flat panel display processing lines with high throughput and output quality.

4.3.6 Etching

Chemical removal of materials through wet and plasma etching is a well established discipline, and several useful guides are available in the literature [62] [63] and from the suppliers of etching reagents. Organic semiconductors and conductors are easily etched using an oxygen plasma, etch rates are similar to those achieved when etching photoresist. It is also possible to etch or de-dope some organic conductors or semiconductors in a strong base or reducing agent and thereby render them inactive.

Exposure to oxygen and oxidizing plasmas can affect the performance of organic semiconductor materials. Steudel, et al. has shown that etch plasmas can affect transistor material and shift the threshold voltage approximately $100\mu m$ away from the perimeter of the resist, and that the device layout should protect the transistor material by overlapping the resist at least by this amount [64].

4.4 Processing considerations for high crystallinity

For most OFETs processes, a major goal is achieving high crystallinity in the semiconductor and avoiding inadvertent doping of the device. Several considerations to achieve these goals will be mentioned.

4.4.1 Polymer crystallinity

Most polymers suitable for OFETs have a semi-crystalline structure. This structure forms during the last stages of the drying process. It is essential during device fabrication that the solvent evolve slowly from the polymer to leave enough time for the internal material organization to occur; moderate volatility solvents are preferred for this reason. Some deposition processes, such as dip-coating, expose the dried material to a solvent vapor which leaves additional time for the material to reorganize and crystallize more throughly.

4.4.2 Stranski-Krastanov growth of small molecule crystallites

Most vacuum deposited small molecule organic semiconductors crystallize following a Stranski-Krastanov growth pattern [56], schematically shown in Fig. 4.9. The pi electron cloud of many small molecule organic semiconductor materials is exposed, and to obtain the minimum potential energy configuration the molecule tries to arrange the cloud in the region of highest polarizability. When oligomers are deposited on an insulator, molecules stand up so that the pi electron clouds sit near those of the neighboring molecules and experience the van der Waals interaction. Each molecule twists to accomodate the slight repulsion between electron clouds, and so that and the slightly positively charged hydrogen ringed backbones face the slightly negatively charged cores of the neighboring molecules. On the electrodes the maximum polarizability is achieved by facing down into the metal, which has significant implications for crystal growth and performance at the channel edges (see Section 5.2.2).

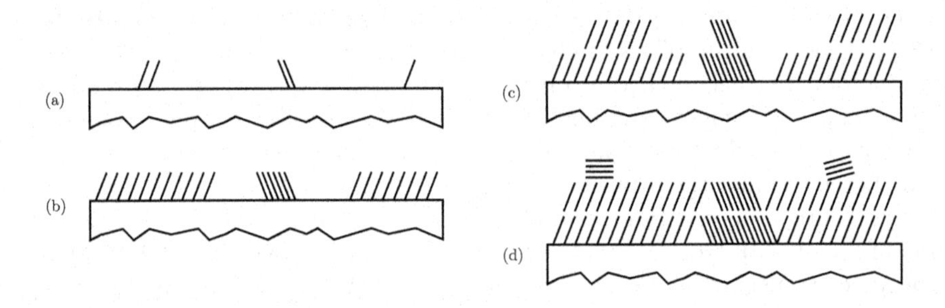

Fig. 4.9. Schematic view of the Stranski-Krastanov growth process. The inital seed layer (a) has a strong repulsion with the substrate and is well organized. (b) As additional material is added it continues to expand the initial seed crystal domains. (c) A second layer nucleates when a certain level of surface coverage is reached, and the first layer continues to grow. (d) Eventually enough material deposits so that the repulsion with the substrate is weaker and the deposited material is no longer organized with respect to the substrate.

During deposition, there will be an optimum substrate temperature (or possibly a series of temperatures as the deposition progresses) for the formation of high quality crystals, exclusion of water, and adjustment of the crystal phase. These parameters will not be exactly the same for each deposition setup, but some typical starting points are presented in Appendix B and in the literature (see, for example, [40] and [65]). Some experimentation is required to determine the optimum thickness, rate, and substrate temperature for a given deposition and material system.

The deposition rate has a clear effect on device performance, but there is some substantial disagreement in the literature on the optimum rates for each

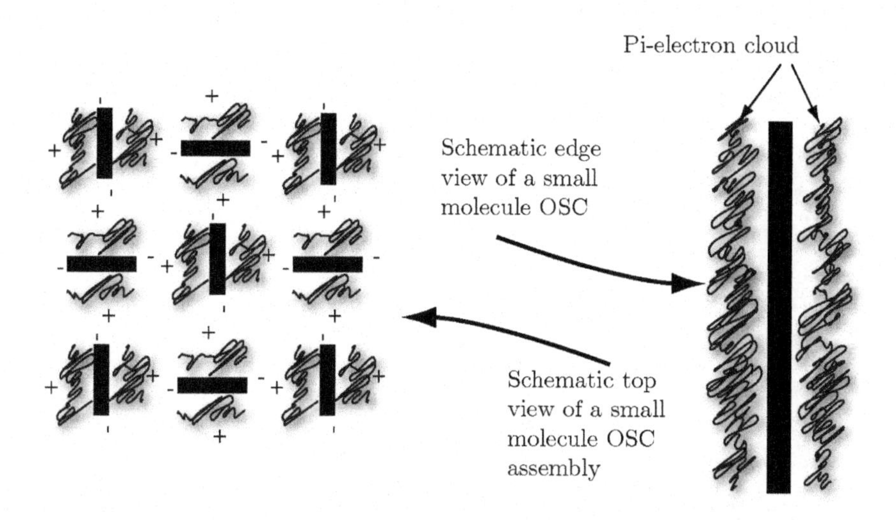

Fig. 4.10. Schematic side and top view of the electron cloud in a small molecule electron-rich organic semiconductor (e.g. an oligothophene or an acene). Because hydrogen is more electronegative than carbon the peripheral hydrogens attract the pi electrons leading to a herringbone stacking. The higher permittivity of neighboring standing molecules leads to the nearly vertical orientation for the first few deposited layers.

material. The deposition rate changes the exposure time of the substrate and deposited material to IR radiation from the evaporation source. This causes a change in the substrate temperature which depends on the rate, evaporation source, and chamber geometry. The kinetics of Stranski-Krastanov surface/island nucleation is also affected by rate. Deposition that is too rapid can constrain surface mobility and lock in a suboptimal configuration, deposition that is too slow can nucleate small islands and reduce the overall crystallinity. The optimum thickness of organic semiconductor material is also a nontrivial issue. Too little material deposited leads to a device which does not conduct, but too much material leads to the formation of parasitic parallel conduction paths in the device which degrades the output resistance. As with the other parameters, the exact pattern of crystallization (which depends on the rate and temperature), growth at the contacts, chamber geometry, and deposition thickness uniformity make it difficult to establish an optimum thickness for each material without experimentation. Coverage experiments have shown that at least the first two monolayers are active in the electrical performance of pentacene and sexithienyl OFETs [66] [67], setting a lower thickness limit for top contact devices. Contact coverage is poor in bottom contact devices at these thin layers, however, and additional material which is deposited further improves the contact behavior.

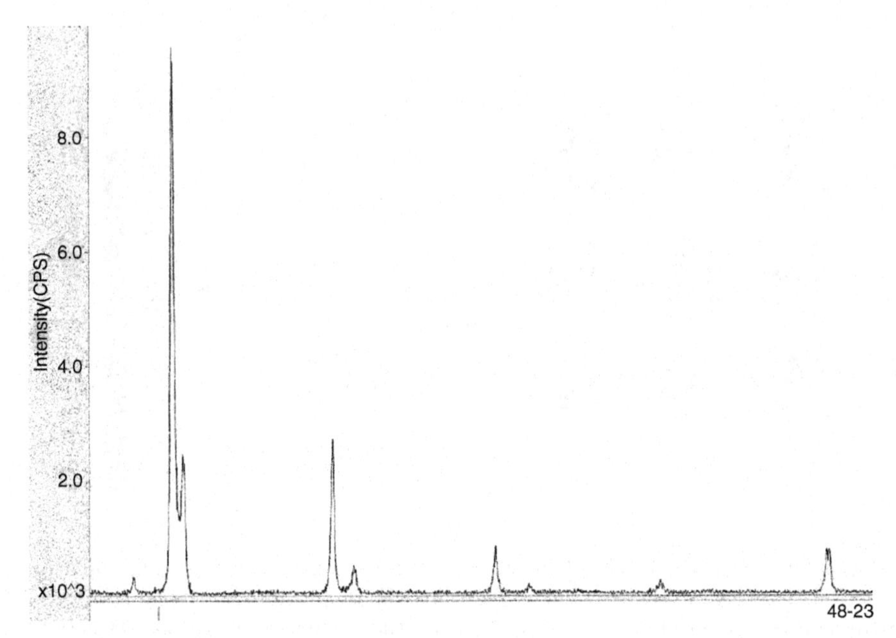

Fig. 4.11. A powder X-ray diffraction pattern taken of pentacene grown at room temperature. The polycrystalline structure is evident, as is the polymorphic structure of the film. Two sets of peaks form corresponding to the thin film phase, which is thermodynamically metastable, and the bulk phase, which is thermodynamically stable. Growth at different temperatures near room temperature can influence the ratio of these two phases through manipulation of the growth kinetics [56].

An interesting byproduct of Stranski-Krastanov growth is that polycrystalline oligomeric semiconductors almost universally need to be bottom gated. The top surface is poorly ordered and exhibits performance orders of magnitude worse than the bottom surface. Many amorphous small molecules and most polymers can be operated with a top or bottom gate.

Solution deposited small molecules experience many of the same organizational issues as soluble polymeric materials. Drying time and environment, interaction with the substrate, and the presence of topography and chemical states on the gate dielectric can all complicate proper crystallization.

A further complication of the Stranski-Krastanov process is that evidence exists that the crystallite size observed through atomic force microscopy may not be directly indicative of the degree of crystal organization in the first few monolayers. As an example, when polycrystalline films are viewed under crossed polarizers, domains significantly larger than those observed through atomic force microscopy are observed. Wang has shown that AFM analysis understates the grain size as measured by crossed polarizers by a factor of 4 [68].

4.4.3 Threshold voltage and bias stress

During processing there are a number of opportunities to create shallow states in the semiconductor which can degrade the device transconductance and shift the threshold voltage through the formation of a charged interface dipoles or bulk trap states and release of mobile charges.

Water incorporation in the semiconductor and gate dielectric is a significant contributor to threshold voltage variation. Water can react with pentacene to form shallow states [69] and can also move in hygroscopic gate dielectrics leading to a long effective hysteresis in the capactiance response [70]. It is possible to reduce the impact of water incorporation by vacuum exposure [71], encapsulating OFETs [72], and using non-hygroscopic materials for gate insulators [70].

An additional opportunity for device doping exists at the gate dielectric-semiconductor interface. It is essential that any process chemicals (e.g. from photolithography) are removed, that unreacted crosslinking agents are passivated, and that polymeric gate dielectrics are protected from oxidizing plasmas from etching, cleaning, and sputtering processes. In addition to directly creating states, dangling bonds on the gate dielectric can also cause significant changes in morphology (see, for example, [73]), which can also affect the charge transport performance.

4.5 Several archetypical process flows and variants

Several archetypical process flows are shown here, primarily based on processes in the literature. All of these provide independent patterning on four layers. A virtually infinite variety of processes can be imagined which combine these or other unit operations, depending on the desired device characteristics, required circuit topography, available equipment, etc. Chapter B provides some tested recipes which can be used as a starting point for development of processes based on a subset of these unit operations.

For the sake of consistency, the four basic OFET layers will be referred to as the gate layer, gate dielectric layer, source/drain layer, and active layer. OFETs with integrated devices may also re-use one of these four layers for hybrid device integration or introduce additional layers for connectivity or other active device formation. The position of these layers is shown in two idealized OFET cross sections in Fig. 4.12.

The gate layer (which is also typically used for some interconnect) is a conductor and is patterned both to establish interconnects and decrease the overlap capacitance with the source/drain layer which is a drag on performance in many applications. The gate layer is typically the best bonded to the substrate and is often also used to anchor layers that will be used for external connection (e.g. through heat seal connectors or wire bonding). The gate dielectric layer is typically patterned to allow interconnection between

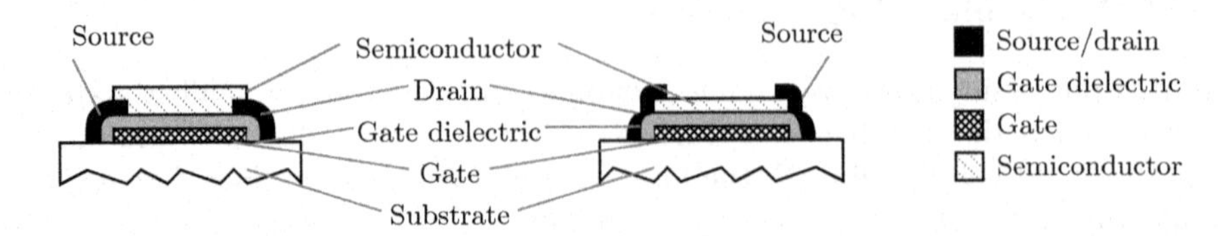

Fig. 4.12. The basic idealized OFET strucure in a bottom gate top and bottom contact configuration. In the processes presented in this section, all four of these layers are patterned. Many, but not all, organic semiconductors can also be fabricated in a top gate configuration.

the source/drain layer and the gate layer, although useful circuit geometries exist in which this is not necessary (e.g. some LCD capacitave drive topologies) and patterning may be skipped. The source/drain layer contacts the semiconductor and is both patterned to establish the channel and also to be used for circuit interconnection. Resistors, vias, and capacitors can be formed by appropriate patterning of these three layers.

The active layer contains the semiconductor. The active layer is typically patterned to avoid leakage currents between transistors, avoid the formation of unintentional parasitic transistors and MOS capacitors, and to avoid unintentional parasitic paths in ungated areas. When the transistor threshold voltage is such that the semiconductor is depleted it is possible in many cases to avoid patterning the active layer. There are also circuit geometries and layout topographies which suppress stray currents and allow the use of unpatterned semiconductor material if that is desirable. These are discussed in further detail in Chapter 5.

4.5.1 Shadow masking

A pentacene OFET process is shown in Fig. 4.13, based on the processes proposed by Kelley and Baude, et al. in [74], [75] and [40]. All four layers in the device are shadow masked, and a passivation layer is used to quench reactive sites on the e-beam thermally deposited Al_2O_3 layer, which has a high density of dangling bonds.

There are several interesting features of this process flow.

The gate dielectric is a thermally evaporated Al_2O_3 layer. Evaporating the material using an e-beam source allows shadow masking, but also deposits a material with a large surface energy state density which complicates nucleation and assembly of crystals in the channel. The authors showed that spin casting a dilute layer of polystyrene or poly-α-methyl styrene passivates these surface states and increases OFET performance without creating a significant insulating barrier between the gate and source/drain layer. This allowed the

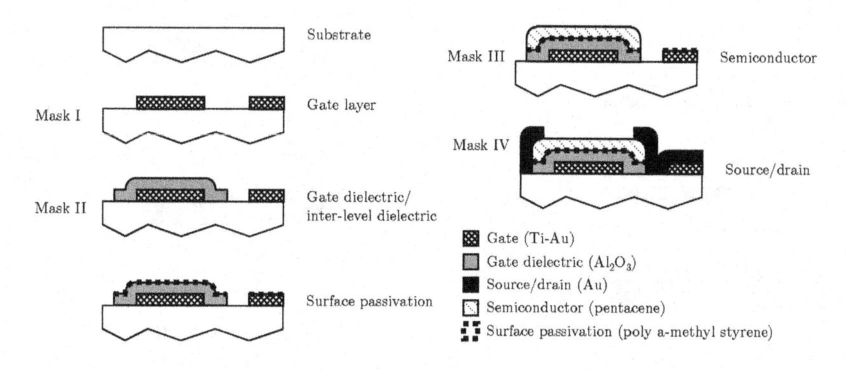

Fig. 4.13. The process flow presented in [74]. Each layer is patterned using a shadow mask, except for the surface passivation layer which is blanket deposited. The surface passivation layer does not introduce any significant resistance between the gate and source/drain layer. The gate dielectric layer is only deposited where necessary due to the nature of the shadow mask geometry.

use of the styrene-based blanket cast film to be used as the surface treatment without the need for any patterning.

Use of shadow masking for the source and drain layer allows for top contact devices to be fabricated, which improves the contact resistance and insures favorable transistor crystal growth in the channel.

Because shadow masks need to be self supporting, the geometries which can be fabricated in this way are limited. For example, it is not possible to create a circular electrode trace in a single masking step. In sparse circuit geometries it is usually possible to arrange circuit traces so that long or circular runs are shared between the two metal layers. For denser or overlapping runs, multiple masks can be used to create the appropriate patterns. Depending on the shadow mask design used, it may also be favorable to split patterns among the gate and source/drain layers. Because there are only two metal layers, only a relatively small fraction of the circuit area needs dielectric material deposited and the shadow mask layout is generally straightforward. An additional advantage of using shadow masks is that there are no residues which need to be cleaned after each step is deposited; layers emerge both patterned and clean after vacuum deposition.

4.5.2 Parylene encapsulation

A subtractive process using pentacene as the semiconductor material based on [76] is shown in Fig. 4.14. The process uses parylene as the gate dielectric and the encapsulation layer for a semiconductor layer using pentacene. The encapsulation layer process protects the pentacene against exposure to the solvents and other reagents used in the photolithographic process.

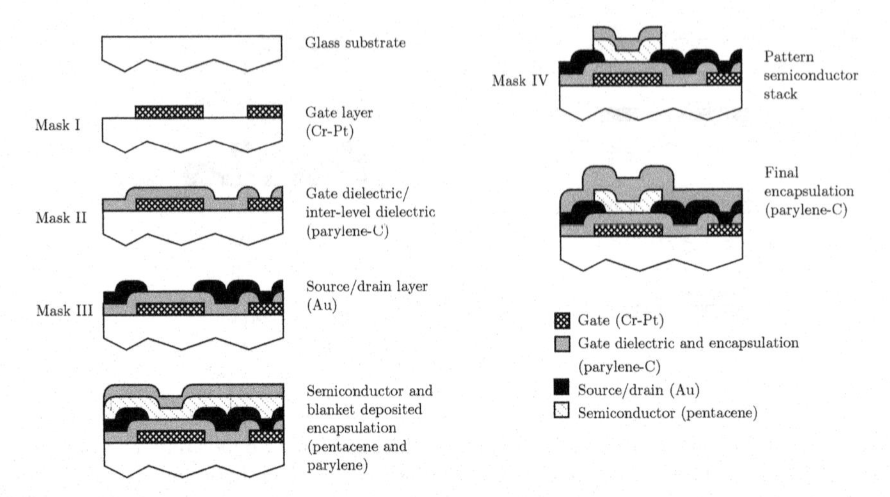

Fig. 4.14. The process flow in Kymissis, et al. [76]. All layers are subtractively etched, and parylene is used to protect the pentacene from attack by the photoresist and developer.

This process is straightforward to implement photolithographically, and can be used with either substractive etching of the metal layers or lift-off. This allows photopatterning and subsequent etching using an oxygen containing plasma.

As with any subtractive processing, photoresist and etch residue cleaning between layer depositions can significantly affect the performance of subsequently deposited layers. Use of parylene as a gate dielectric and/or encapsulation layer has been applied to a range of organic semiconductor materials which have some solvent or plasma sensitivity, including semiconducting polymers as well as small molecules [77].

4.5.3 PVA resist

An alternative to encapsulating OFETs with parylene is to use an aqueous photoresist system which does not induce the destructive phase transformation which is observed when metastable oligomeric semiconductors are exposed to other solvents. Since the resist is developed in pure water, exposure to strong bases or solvent developers is also avoided. The process shown is based on Kane, et al. [57].

A unique advantage of using a semiconductor-compatible resist is that the surface can remain available for further device processing. The same team has also adapted the process to a top contact process flow with improved contact resistance performance [78].

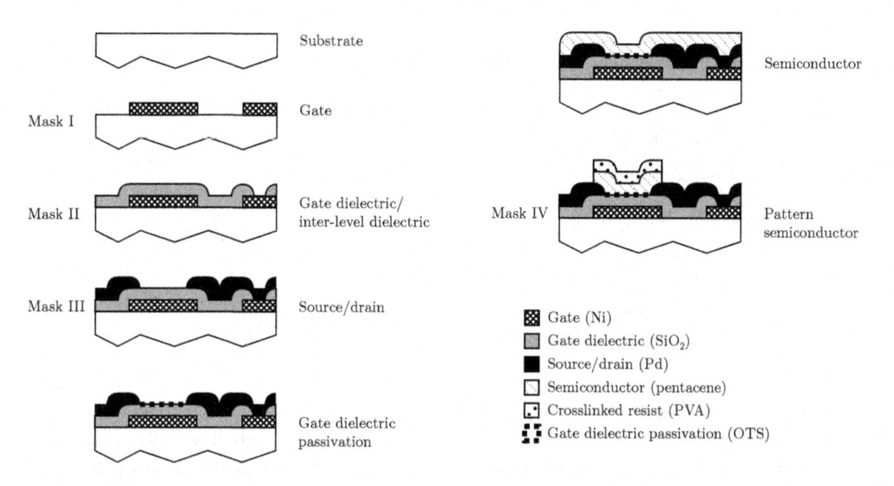

Fig. 4.15. The process flow described in [57]. The gate, gate dielectric, and source/drain layers are photolithographically patterned. The gate dielectric is passivated using octadecyltrichlorosilane, and the semiconductor is deposited. A chromate sensitized aqueous polyvinyl alcohol based photoresist is then used to pattern the semiconductor. Uncrosslinked resist is developed with water and the exposed pentacene is etched using an oxygen plasma.

4.5.4 Subtractive inkjet/digital lithography

A significant challenge in inkjet printing is the formation of the gate dielectric layer and smooth gates. Several hybrid inkjet printing processes have been demonstrated which use photolithography of one or more layers to overcome this obstacle, progress continues to be made on all inkjet printed devices. The wetting/dewetting dynamics of printed materials at the surface and the thermodynamics of the drying process complicate the fabrication of smooth and flat films. Surface roughness at the gate dielectric interface decreases device transconductance (likely through a combination of increased path length, frustrated crystallization, and surface scattering), and pinholes can cause fatal leakage in devices.

A solution to this problem, which retains many of the advantages of printing, has been demonstrated by Arias et al. [79]. In the demonstrated process, blanket films for the gate, gate dielecric, and source/drain layer are deposited and a protective wax resist is dispensed using jet printing. The layers are patterned and the resist is cleaned from the substrate. Because the dielectric is deposited using PECVD at relatively high temperatures (350°C), it is of higher quality than achievable using printing. The process retains the digital control offered by inkjet printing. The semiconductor used is PQT12 which is a soluble semiconducting polymer. It is additively printed after the gate dielectric surface is treated with octadecyltrichlorosilane.

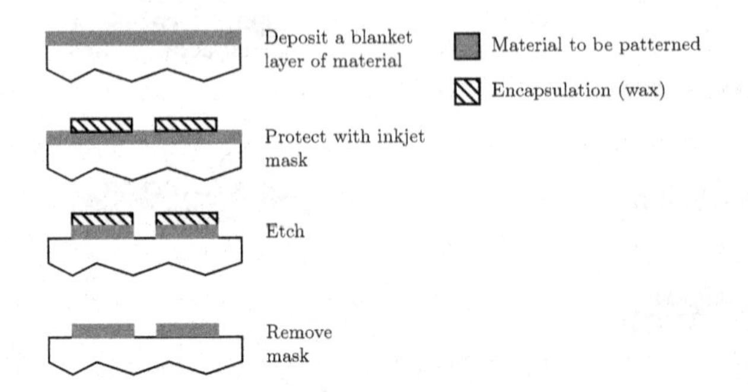

Fig. 4.16. A schematic showing the basic concept of subtractive inkjet that the authors have named digital lithography [79]. This process patterns using a digitally controlled method, but allows for deposition using any technique. The material to be patterned is first deposited. A molten wax is jetted in the desired pattern. The wax solidifies on contact with the surface. The unprotected areas are etched using conventional processes, the resist is stripped, and the next layer processed.

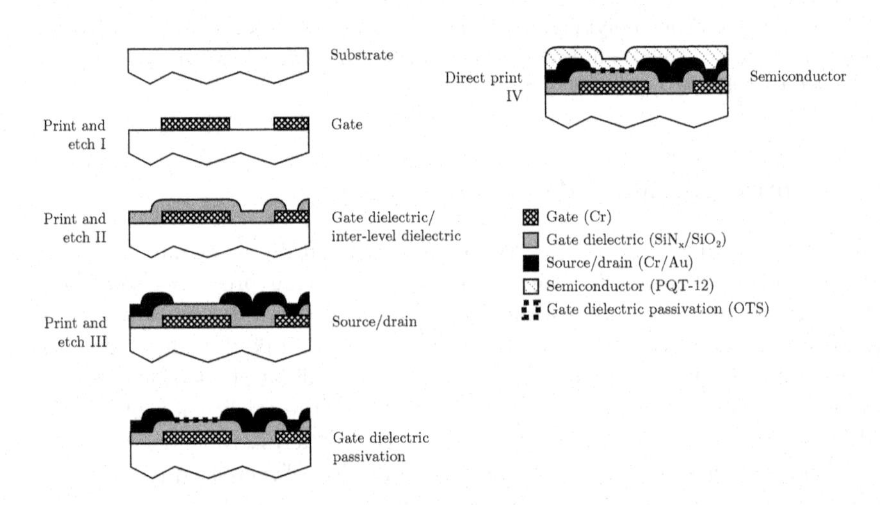

Fig. 4.17. The process flow used in [79]. The first three layers are deposited using conventional blanket deposition techniques and are patterned using the digital lithography process shown in Fig. 4.16. The metal layers are thermally evaporated and the nitride/oxide composite layer is deposited using PECVD. After treatment of the stack with octadecyltrichlorosilane to passivate the surface the semiconductor is deposited additively using a jet printer from solution and dried.

4.6 Conclusions

This section discussed the basic unit operations and process flows for OFET fabrication. Several example processes were presented which integrate these process flows into a system capable of producing an integrated circuit. The next section will explain some additional strategies for improving the performance of these devices.

5

Advanced OFET fabrication

5.1 Introduction

Chapter 4 described many of the fundamental unit operations used to create thin film organic FET structures. A range of unit operations and process flows can be used to take advantage of the van der Waals bonded structure of organic semiconductors and create integrated circuits with a range of functionalities. This chapter will explain some performance related processing issues which have been identified, along with some suggestions for more advanced processes with higher performance.

5.2 Source and drain contacts

One of the critical operating junctions in OFET devices is the source and drain contact to the channel. Any barrier between the contacts and the channel will appear in series and impede the flow of charge through the device. The source and drain electrode formation and structure can also influence the properties of the transistor channel itself. Crystal growth nucleated on the source and drain and the processes used to pattern the source and drain electrodes can have a significant effect on overall device performance.

5.2.1 Work function considerations

An additional consideration in the selection of source and drain electrodes is the work function difference between the semiconductor and the energy levels in the semiconductor. Gold is a popular choice for PFETs, since its large work function provides reasonable access to the HOMO level of many materials. There are some indications that noble metals like Pt with a higher work function (and, in principle at least, a better energy match) may provide even better contact performance. It is not entirely clear, however, whether these

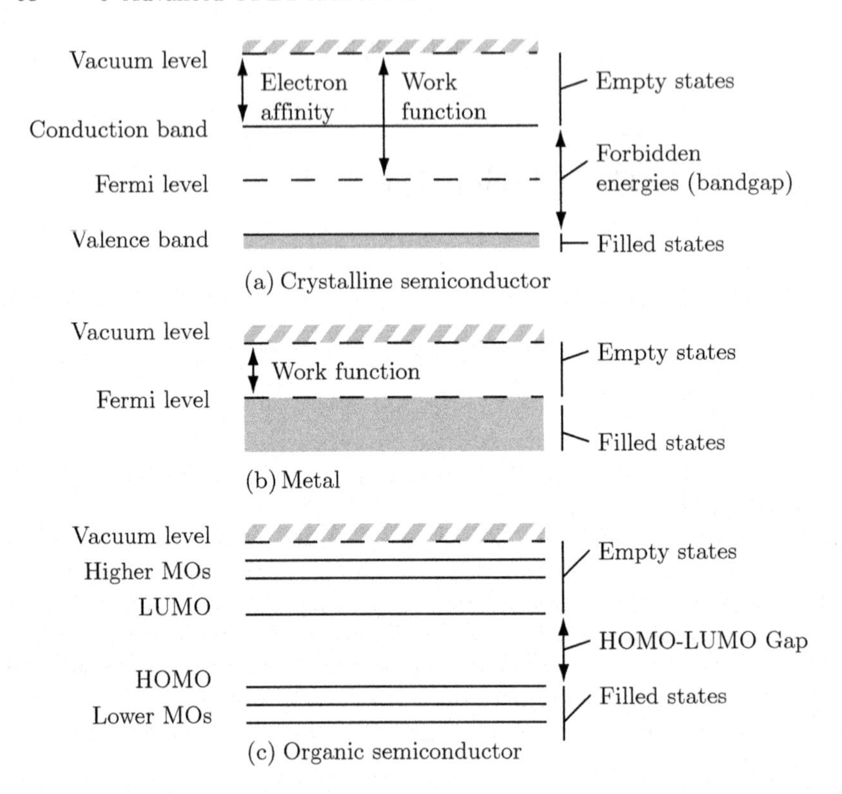

Fig. 5.1. A schematic diagram of the energy levels and filled/empty states for (a) a band-transport semiconductor, (b) a metal, and (c) an organic semiconductor in the absence of thermal excitation or doping. In crystalline and organic semiconductors, an energy gap exists, whereas for a metal, energy levels immediately above the filled states are permitted. Note that carriers ejected beyond the vacuum level can be at any energy.

effects can be neatly divorced from morphological, charge transfer, surface dipole, and other second order effects which also influence the contacts [80].

In principle, it should be favorable to use low work function contacts for n-type organic semiconductors. Low work function materials will sit higher in the energy level diagram and will have superior access to the LUMO. Many investigators have found, however, that performance with low work function materials is inferior to that achieved using noble metals. A leading hypothesis is that the insulating surface oxides found on these materials interferes with conduction in many of these systems. Low work function materials are naturally reactive, because the activation energy for removing an electron is low, allowing for easy oxidation reactions involving water and oxygen. A surprising outcome has been that gold appears to be the best choice for many n-type semiconductor materials not because of any particularly good energy

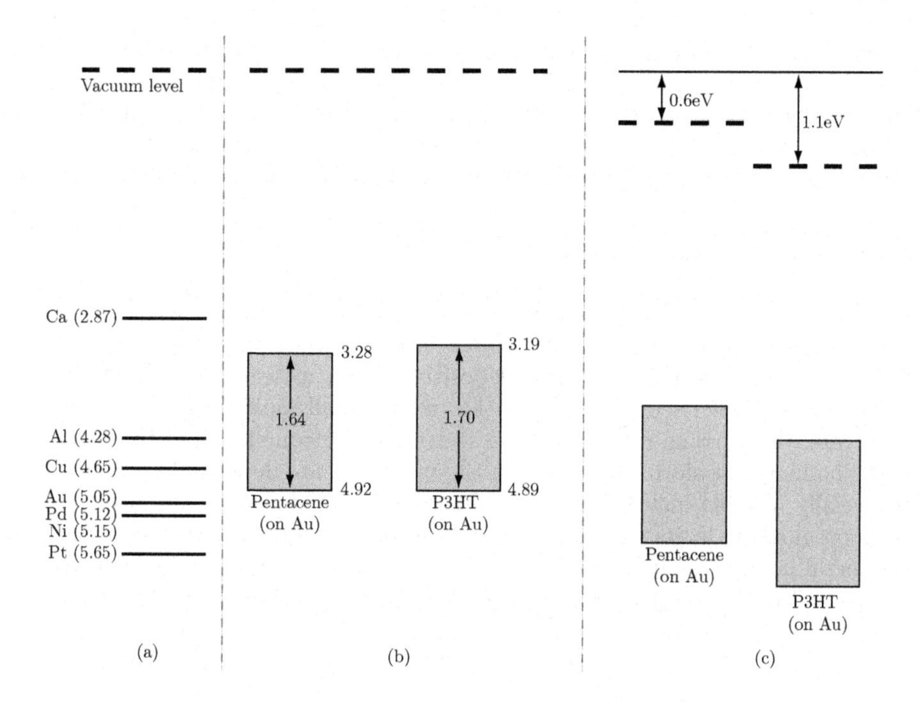

Fig. 5.2. A chart showing the work function of several metals (a) [81] and the HOMO/LUMO levels of several organic semiconductors relative to the vacuum level (b) [82] [83] in eV. Theoretically, in n-type devices we want to inject electrons and therefore would expect to see the best performance using electrodes whose work function lines up with the LUMO, and for p-type contacts (injecting holes) lined up with the the HOMO. In practice, noble metals appear to work best for most material systems. (c) Contact dipoles caused by chemical interactions with the metal surface, dopants, and contaminants can vary the energy level/contact offsets–those seen by [82] [83] are shown. On this diagram, these offsets appear as a shift in the vacuum level. Note that the effective energy levels for organic semiconductors will differ on different surfaces; these are both measured deposited in a thin layer on gold.

level match, but because of its air stability and lack of any native oxide to form an injection barrier [17].

Several alternatives are under serious consideration in the quest for better energy level matching and contact injection. Conducting metal oxides, which offer a wide range of energy level tunability, are promising candidates [84]. Doped organic semiconductors (such as polyaniline and PEDOT) offer a reasonable energy level match and can potentially either locally dope the semiconductor with the contact dopant and offer a better surface for oligomer organization than metals [85]. The damage layer formed when electrodes are evaporated on top of organic semiconductors may offer a better contact, especially for n-type materials (this is the case in OLEDs). Finally, the use of contacts which contain nanowires or carbon nanotubes also appears to improve injection into the channel considerably [86]. The superior contact behavior in these materials appears to be both a function of the energy level match and enhanced injection due to field enhancement, which concentrates field lines at the contact edges and enhances injection over any remaining barrier.

5.2.2 Top vs. bottom contacts

In addition to any enhancement caused by local damage during evaporation, the structure of the crystal at the source/drain interface can be complicated by the presence of a metal when depositing small molecule semiconductor materials. As discussed in Section 4.4.2, when small molecules land on the substrate, they arrange themselves to seat their electron clouds in the region with the highest polarizability. When the molecules are deposited on a metal (especially a noble metal, like gold, which has no oxide on the surface), the electron cloud is in its lowest potential energy configuration facing flat into the metal layer. This makes small molecules lie down on the substrate at the source and drain contacts, in a different configuration than that observed in the channel (where the substrate is insulating) [87] (see Fig. 5.3).

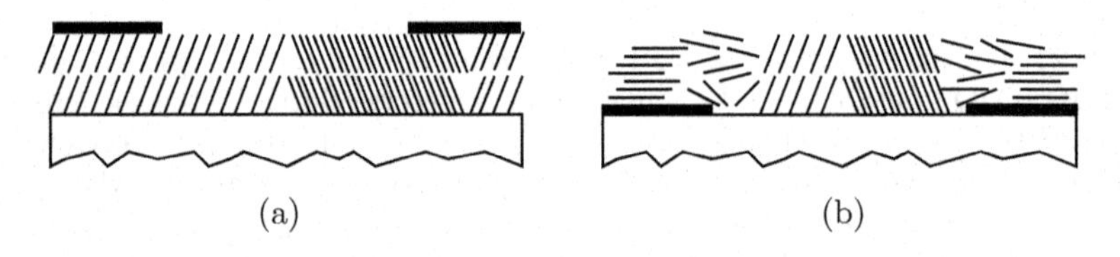

Fig. 5.3. A schematic illustration of the molecular packing in (a) top and (b) bottom contact OFET devices. Small molecules deposited on metal contacts tend to lie flat, and do not have a contiguous grain structure where the molecules are standing vertically at the center of the channel.

This growth pattern leads to a crystal frustration region that reduces device performance in bottom contact configurations due to both a larger series resistance and degraded performance in the section of the channel which has an inferior growth pattern. Fig. 5.4 shows scanning electron micrographs of the crystal grain structure at the channel edge and the resultant frustration region. This phenomenon has been observed both in vacuum deposited [87] and solution deposited [88] materials. If the contacts are undercut the effect is even more pronounced [89].

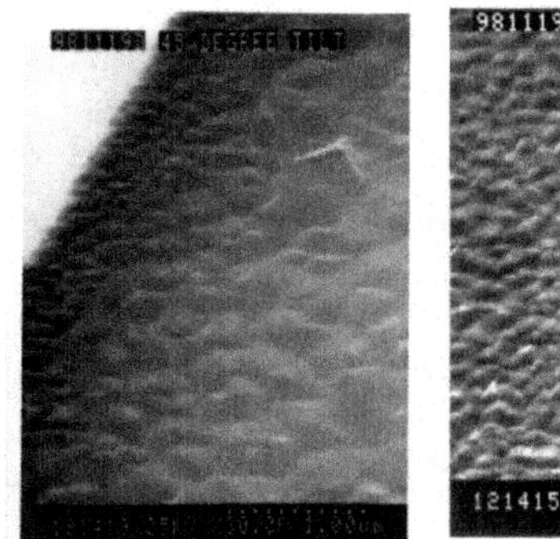

(a) Untreated electrode edge (upper left is gold electrode)

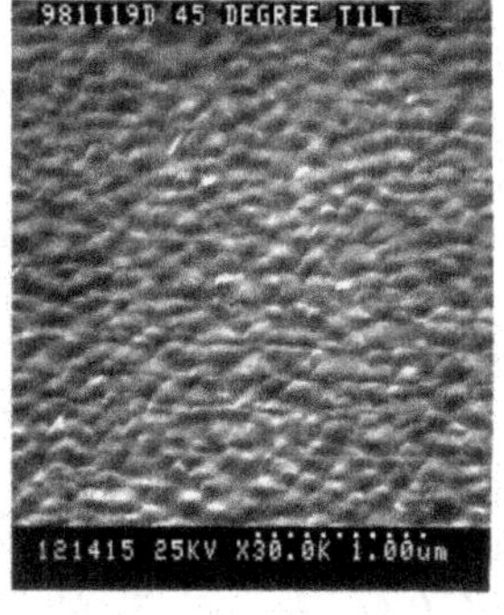

(b) Untreated electrode

(c) Treated electrode

Fig. 5.4. Scanning electron micrographs showing the packing of pentacene near treated and untreated gold source/drain electrodes. (a) shows the untreated gold electrode edge; the transition from the electrode influenced structure (b) to the larger grain structure in the center of the channel can be clearly seen. (c) shows a thiol treated electrode; the grain structure matches and several grains are contiguous across the electrode edge.

5.2.3 Treatment of contacts

Many process flows require a bottom contact electrode configuration. For these processes one solution to the bottom contact problem is to treat the contacts to improve the device behavior.

One method to engineer the condition of surfaces selectively is to use self assembled monolayers (SAMs). SAMs generally have a head which selectively attaches to a material and a tail which is inert against the head. Once the active sites on the target substrate surface are all occupied, the SAM stops attaching and under ideal conditions a single monolayer forms which can be very tightly packed and highly ordered. The surface then acquires the properties of the SAM tail with this single molecule thick coating. Even imperfect SAMs can exhibit dramatic changes in surface energy and condition. A range of SAMs have been developed with a range of functionalities (e.g. attachment selectivity, cross-linking, hydrophobic or hydrophilic character, amine-termination, fluorination, selective tail reactivity, etc.).

One major class of SAMs is the thiols, whose general structure is shown in Fig. 5.5 (a). Thiols attach to gold, platinum, and a number of other metal surfaces. By treating the surface of the source and drain electrodes with a thiol, it is possible to create a low surface energy layer on the source and drain which is not strongly insulating. Oligomeric semiconductors can grow on this surface with the same structure as an insulator. Fig. 5.4 (c) shows

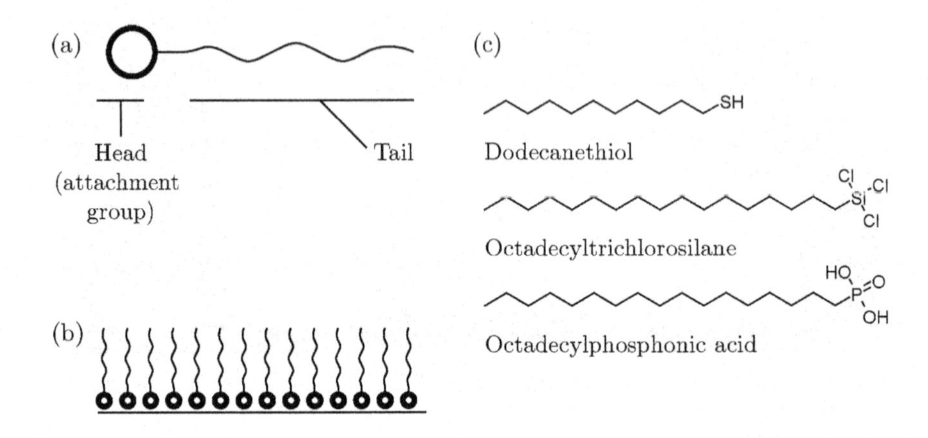

Fig. 5.5. (a) shows the general structure of a SAM, (b) schematically shows the packing on a substrate, and (c) shows several representative SAM structures

pentacene growth at the edge between a SAM treated electrode and an insulating transistor channel; contiguous grains can be seen growing across the edge. The performance of devices with these improved contacts is significantly better than those with untreated contacts.

The top/bottom contact behavior seen in small molecules is also seen to a certain degree in solution deposited materials as well. An appropriate surface energy is required to seed proper assembly. Just as poor growth on the contacts extends some distance into the channel, so does favorable growth. As an example, Gundlach, et al. developed a platform in which spontaneous assembly on the gate dielectric resulted in poor performance, but the favorable growth on SAM treated source/drain electrodes extended many microns away from the electrodes, bridging channels in many cases [88]. The difference between the transistor performance near the electrodes and some distance away was large enough that removing the semiconductor was unnecessary for creating integrated circuits. Taken to the extreme, tuning the surface energy for optimal growth can nucleate and create selectively grown patterned single crystal devices [90].

There is evidence that the treatment of the source and drain contacts can be used to create a dipole that assists in the injection of carriers. It is well established that electronegative self assembled monolayers can sigificantly increase the effective work function of a noble metal surface, which should improve the injection into PFETs. It is not entirely clear how best to separate work function from structural effects, but studies done on progressively electronegative SAM series show a significant improvement in contact behavior and linear region behavior which is correlated with the electronegativity of the SAM, consistent with this hypothesis [91]. This effect is analogous to contact doping in a-Si FETs, since it creates a region with high charge carrier density at the contacts only that facilitates charge transport. The use of doped and undoped organic semiconductors (e.g. PEDOT or MTDATA) is also effective [92]

For both crystal texture improvement and work function engineering SAMs are particularly favorable and straightforward to handle. When deposited under the right conditions, they selectively attach only to the electrodes and self-limit their growth to one monolayer. This allows for straightforward deposition without the need for any additional lateral patterning or precise application to limit the layer's thickness.

5.2.4 Creation of lithographic top contact devices

Most lithographic processes require bottom contact devices because the semiconductors used cannot tolerate exposure to the process chemicals which would be required to create top source/drain contacts. The development of photoresist materials and processes which use aqueous [57] or fluorinated [59] solvents which can directly contact organic semiconductors without degrading their performance has relaxed this constraint. It is now possible to lithographically define top contacts on material which has been deposited on a favorable dielectric for semiconductor growth and achieves superior device performance [78]. Top contacts may also be advantageous because of the improved contact through the damage layer created when the metal condenses on the surface.

It is also possible to create top contact devices using printing processes. Transfer and contact printing has been explored by several groups. Deposition of electrodes from solution on pentacene and related polymorphic materials was once thought impossible because of the solvent phase conversion problem. Using an advanced femtoliter inkjet system in which the solution dries before hitting the substrate, however, Sekitani et al. have been able to directly print top contact devices [93].

5.3 Gate dielectrics

As with the source and drain electrodes, the composition and condition of the gate dielectric plays an important role in the performance of OFET devices and can also affect their functionality.

5.3.1 Characteristics of gate dielectrics

A large variety of materials have been used as OFET gate dielectrics. A number of considerations typically go into the process of gate dielectric selection including:

- Leakage

- Patterning convenience and compatibility
- Semiconductor compatibility
- Achievable capacitance
- Surface tunability
- Hysteresis

Hysteresis in OFETs can arise from a number of sources (this is further discussed in Section 6.2.4), one of which is water incorporation in the gate dielectric material [94]. Organic gate dielectrics with strong polar groups can absorb water and experience a slow relaxation which can compromise device performance. When a gate dielectric is selected, attention should be paid to either eliminating this issue through selection of a less hygroscopic dielectric material (e.g. an inorganic or hydrophobic material), or by dehydrating and encapsulating the device.

5.3.2 Crystal structure improvement

In bottom gate devices, in addition to insulating the channel the gate dielectric is also the surface which templates appropriate growth for the semiconductor material. It is possible to improve the growth of organic semiconductors on gate insulator materials through appropriate surface modification. In a top gate device the substrate surface can be engineered as well as the interface with the subsequently applied gate dielectric.

Perhaps the best studied gate dielectric treatment is the application of silanes (e.g. octadecyltrichlorosilane or OTS) to thermally grown silicon dioxide. While this structure is limited in its utility in integrated circuits, it provides a well controlled system in which to study structure/function relationships that can then be extrapolated to more practical, but less controlled, integrated structures (e.g. PECVD or PVD deposited oxides).

It is clear that the application of silanes to SiO_2 significantly improves the performance of a range of small molecule and polymer semiconductor materials [17]. A number of hypotheses have been advanced as to why including passivation of dangling bonds, higher surface mobility during assembly, better solvent wetting from solution, and improvement of the source/drain electrode edge morphology. While silanes normally are expected to attach to silicates and some other material groups, it has been observed that silanes can passivate a variety of dangling bonds and can modify the behavior of a range of surfaces.

Another self assembled monolayer which has been applied to OFETs is based on phosphonic acids. Phosphonic acids attach to many metals and metal oxides, including alumina, and provide essentially the same passivation and surface energy engineering on that dielectric system [40].

Another approach to surface passivation is to coat the surface with trivially thin layers of organic materials which passivate surface dangling bonds. It has been shown that even a very dilute treatment with polystyrene or polystyrene derivatives can have a dramatic effect on semiconductor growth on disordered gate dielectric materials [74]. Cycloaddition of cyclohexene onto surfaces has also been used for surface passivation. Hergendorf et al. showed that a significant fraction of the first pentacene monolayer on clean oxide deposits face-down, presumably reacting with dangling bonds on the surface [73]. He further showed that a surface treated with cyclohexene starts growing crystalline pentacene immediately and has significantly larger crystalline domains than the unpassivated surface.

5.3.3 SAM gate dielectrics

Another application of SAMs to OFETs is the use of insulating high quality SAMs as gate dielectrics. Because SAMs are only one molecule thick and self limiting, it is possible to create almost perfect, low leakgage insulators with high capacitance through careful engineering of the surface condition and application of the SAM. This high gate capacitance makes it possible to achieve very high tranconductance devices with low leakage and under very mild conditions of gate dielectric deposition. Several high performance integrated circuits have been demonstrated using this concept [95].

5.3.4 Introduction of surface dipoles

It is possible to engineer surface dipoles at the gate dielectric interface to manipulate the charge carrier density in the channel and change the effective threshold voltage of the device. This is achieved by placing an electronegative or electropositive material at the gate dielectric surface which donates or withdraws a charge from the device channel.

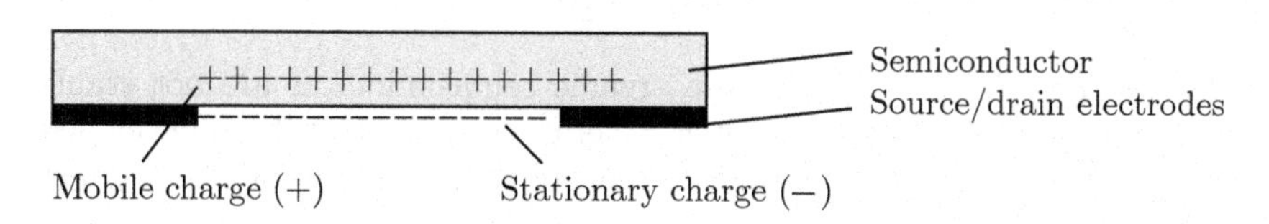

Fig. 5.6. A schematic showing the dipole formed by treatment of the gate dielectric with an electronegative species. Even though the total dipole is electrically neutral, the process releases positive carriers into the channel and shifts the threshold voltage positive. Negative charges can be introduced instead using an electroposive treatment (e.g. evaporating a thin layer of an alkali metal). The shift in threshold voltage can be seen as a capacitor integrating potential over the length of the dipole; no net charge is introduced in this picture.

The basic mechanism for the effective threshold voltage change is shown in Fig. 5.6. Even though the total dipole is electrically neutral, one of the charges is stationary and the other is mobile. The mobile charge contributes to the charge in the channel and shifts the effective threshold voltage by an amount equal to:

$$\Delta V_t = \frac{Q_{mobile}}{C_{ox}} \tag{5.1}$$

Where ΔV_t is the change in the threshold voltage, Q_{mobile} is the mobile charge contributed from the dipole and C_{ox} is the gate dielectric capacitance.

Two possibilities for introducing this dipole are use of an electronegative/electropositive self assembled monolayer [96], and treatment of a polymer gate dielectric with an oxidizing group [68]. The threshold voltage change resulting from an oxidation treatment are shown in Fig. 5.7.

One concern about any such treatment is that the donating or withdrawing group will form a new energy level (or distribution of levels) in the semiconductor. These energy levels may impact the device performance by scattering carriers or degrading the subthreshold slope.

5.3.5 Functional gate dielectrics

Another consideration for gate dielectrics is selecting materials which provide interesting device functionalities.

One functional gate dielectric of interest to a number of groups is using a ferroelectric material to retain state in transistor devices. This has a range of applications including latching circuits [97] and memory storage elements [98].

5.4 Air sensitivity and encapsulation

As discussed in Section 3.2.1, many organic semiconductors are not stable against exposure to oxygen and water. Several strategies have been developed to remove these elements from organic semiconductor layers and seal against their penetration.

Perhaps the simplest strategy is to laminate a plastic sheet on top of the transistors with a suitable adhesive [99]. It is also possible to coat OFETs with a layer of parylene, teflon, or other compatible polymer and optionally seal the devices using a metal vapor barrier layer (e.g. [100]). Heating many OFET materials for dehydration in an inert ambient is also possible, and may be important depending on the material's reactivity with water [101].

The OLED research and development community, which has stringent water and oxygen permeation requirements (significantly greater than those typically considered for OFETs), has several examples which can also be followed. If an impermeable rigid substrate (e.g. glass) is used, a perimeter epoxy seal with a rigid lid (glass or metal) can be used. A dessicant can be placed

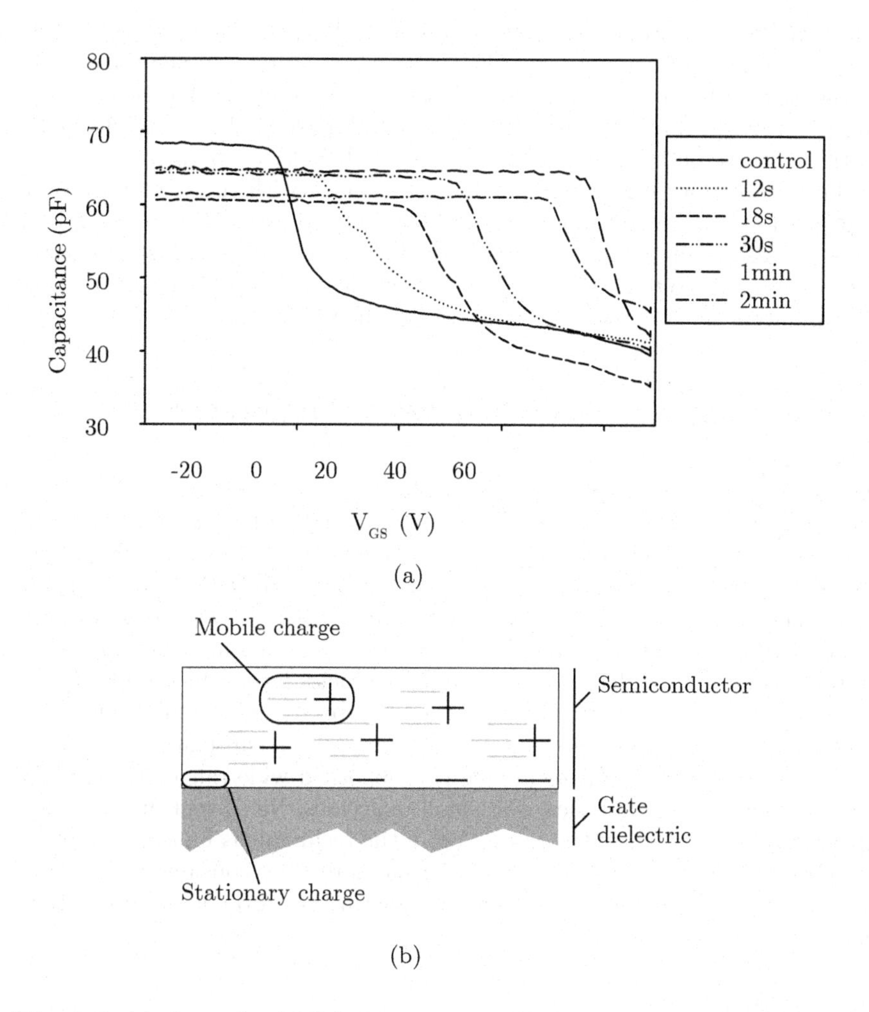

Fig. 5.7. (a) shows the QSCV of a pentacene OFET doped using a surface dipole method [68]. (b) shows the stationary and mobile charges involved in the process. The same general principle also applies to the use of electronegative/electropositive silanes and other SAMs on gate dielectric layers for manipulating the threshold voltage.

in the cavity to continue pumping water after its permeation through the epoxy. A number of low-permeability flexible laminating adhesives are commercially available for laminating together flexible impermeable sheets (e.g. barrier coated plastic foils, thin glass, or stainless steel sheet). While plain plastic may not be adequate as a barrier for OLEDs, its permeability may still be suitable for OFETs without modification or with the deposition of a simple barrier layer such as sputtered ZnO or SiO_2. More sophisticated barrier layers are also available for use with transparent flexible substrates. A comparison of thin film OLED encapsulation strategies with a particular emphasis on transparent flexible barriers can be found in [102].

5.5 Emerging deposition and patterning processes

5.5.1 LITI

Laser induced thermal imaging (LITI) and related techniques deposit patterned films from blanket deposited donor sheets. The donor sheet is placed in proximity to the substrate and a focused laser is directed at areas of the sheet where deposition of the donor material is desired. Depending on the design of the donor sheet layers, the laser can cause thermal evaporation of the target material or decomposition of a sacrificial layer which deposits an overlying material. A range of geometries has been engineered including a range of blackbody absorbers, sacrificial films, source substrates, etc. With appropriate engineering the technique can deposit a wide range of materials including polymers, nanotubes, and small molecules. No solvent or developer exposure is necessary, and the resolution of the technique is limited primarily by the ability to focus the laser and thermal spreading considerations. This approach has been demonstrated on the micron scale over large areas [103] [104].

5.5.2 OVPD

An alternative to thermal evaporation which has been developed is organic vapor phase deposition. In OVPD, an evaporated small molecule is entrained into a controlled inert gas jet from which it is deposited onto the substrate. The process gives excellent control over the morphology and characteristics of the deposited film [48]. This deposition can be performed as a blanket operation using a showerhead or using a small nozzle which can be translated and the gas get pulsed to create a defined device pattern with almost any small molecule material [105].

5.5.3 Surface energy modulation

Solution deposited materials can be patterned by defining areas on the substrate where the solvent wets well and other areas where it wets poorly (e.g.

using self assembled monolayers or pattered thin films with strong repulsive properties). Local islands of the solution are then formed by dipping the substrate into a solution of the semiconductor material, slow spin coating, or another technique which allows wetting and de-wetting to occur leaving patterned areas where the surface affinity is high. When the islands of solution are then dried, depositing patterned films in those areas. This can allow for high throughput pattern definition without the need for printing or subtractive patterning. [106] [107]

5.6 Alternative OFET designs

5.6.1 SIT

The basic transistor current equation law limits the geometrical scaling of current as a function of the width and length (or $\frac{W}{L}$). To increase the transconductance, therefore, either a large W or a small L is desirable. Increasing W consumes device area, whereas decreasing L requires finer and more sophisticated lithography.

The static induction transistor (or SIT), popularized in organic semiconductors by Kudo and colleagues [108], overcomes this bottleneck by sending charge vertically through the device instead of laterally as shown in Fig. 5.8.

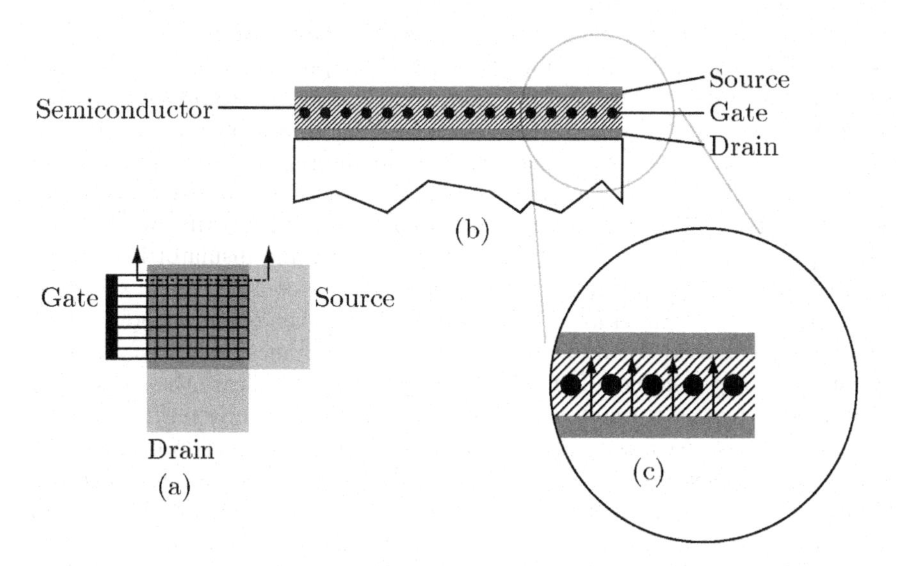

Fig. 5.8. A schematic showing the structure ((a) top and (b) cross-sectional view) and (c) direction of charge movement in a static induction transistor (SIT).

In this situation, current flows through the device's entire area (instead of only in the first monolayer or two on the surface) and the channel length is determined by the total thickness, which is easily controlled in the fabrication stage. Devices made in this manner have a much higher transconductance than comparable lateral devices made using the same semiconductors.

SIT devices necessarily require a process which can fabricate a gate on top of the semiconductor material, which is possible either through selection of the semiconductor, photoresist system, or both. The semiconductors used also need to have reasonable vertical transport characteristics; many amorphous and semi-crystalline semiconductors can be used. Leakage into the gate is of significant concern; selecting a gate material which acts as a Schottky barrier against carrier transport can help reduce gate current in the device.

5.6.2 Reduced patterning processes

A typical integrated OFET process, as described in Chapter 4, takes four patterning steps to complete. The gate, gate insulator, source/drain, and semiconductor layers all need to be patterned.

It is certainly possible to create a range of circuits without patterning the semiconductor layer, and in so doing, save one patterning operation. This is especially possible with logic elements, where it is often possible to achieve adequate noise margin even with some crosstalk, or where space is not of great concern and transistors can be well spaced.

Several architectures have been proposed to reduce the process complexity, especially patterning the semiconductor. One approach is to use a Corbino gate architecture [80]. The Corbino gate, instead of using lateral source/drain elements, uses a nested ring for the source and drain as shown in Fig. 5.9 (a). The outer electrode, the source, acts as a guard and is at the same potential as other sources in the area, eliminating leakage through the unpatterned semiconductor layer. A drawback to the Corbino architecture is that when properly implemented at least one additional metal layer and insulator is required to create the topography resulting in no net mask count savings. Guarded structures can be formed as a compromise, with either reduced performance (through a decreased W and increased overlap capacitance) or increased inter-transistor leakage and no additional patterning. In such a case the source and drain can be bussed through the gate layer to avoid extra patterning.

A second approach is to pattern the gate and the source/drain using the same step, as shown in Fig. 5.10 [109]. A disadvantage of this approach is that there is some ungated transistor (effectively with $V_G = 0$) between the edge of the source and drain and the accumulated channel over the gate. This limits the applicability of this approach to processes where the devices are at least partially accumulated under zero bias.

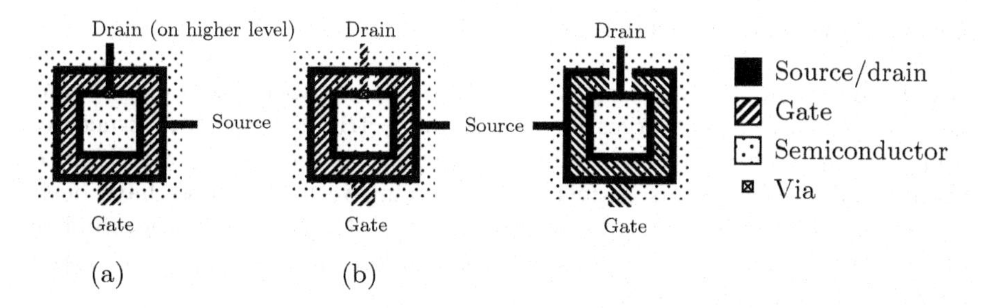

Fig. 5.9. (a) A true Corbino, and (b) two partially shielded drain structures. While the Corbino architecture requires an additional metal layer to feed the drain electrode into the center without shorting the gate, it eliminates the need to pattern the semiconductor due to the source, which shields stray leakage currents. The partially shielded structures are a compromise between the lateral structure and a full Corbino architecture. The performance achieved depends on the layout details of the circuit in question and the ungated sheet resistance of the semiconductor.

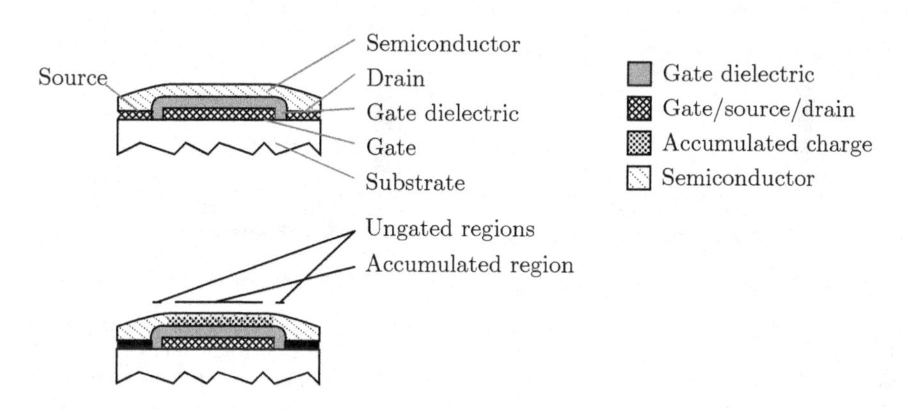

Fig. 5.10. A schematic cross-section of a reduced mask process, which patterns the source, drain, and gate all in one step. The source and drain are exposed during the patterning of the gate dielectric. This process produces a region of ungated semiconductor, and is therefore only applicable to processes which produce depletion mode devices.

5.6.3 Electrochemical OFETs

An alternative approach gating OFETs is to use electrochemical doping and dedpoing to introduce charges instead of an electric field. Such a structure, even though it acts slowly, transduces a chemical potential into a resistance change and exhibits transconductance and gain. Such structures have been of particular interest for use as sensor elements [110] [111]; a large number of selective oxidation/reduction reactions are known which can detect the quantity of an analyte present.

The basic concept behind electrochemical OFETs is shown in Fig. 5.11.

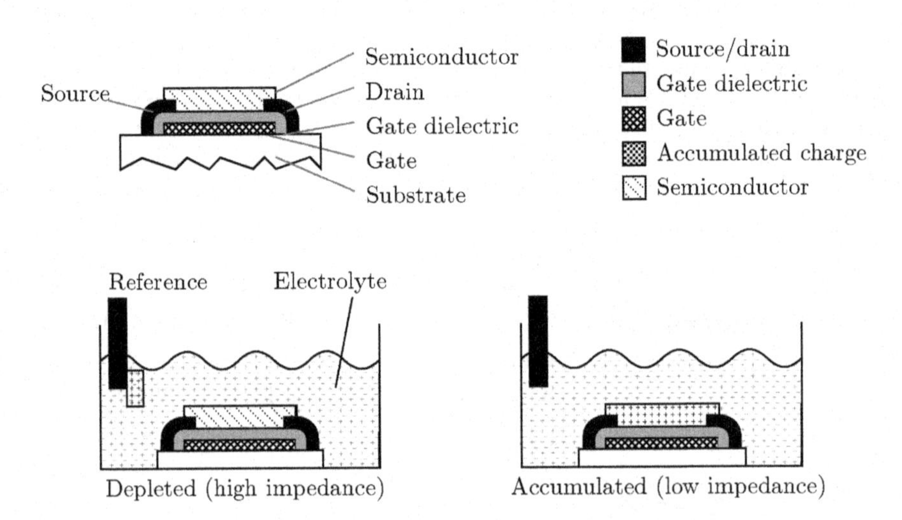

Fig. 5.11. A schematic showing layout of one possible electrochemical transistor (several designs are possible), and the basic principle of operation. In this geometry, charges are moved into the semiconductor from the electrolyte and transistor is turned on. When these charges are biased out of the semiconductor, the device is turned off. The charge accumulation and depletion can be caused by a number of stimuli including the application of potential to the underlying gate, a change in the reference potential, a change in pH of the solution, oxidation/reduction reactions, or chemical reactions on the surface of the semiconductor. The source/drain electrodes can also serve as reference electrodes, and the gate can be capacitavely or directly coupled to the semiconductor.

Dopants present in the electrolyte solution can be driven into or out of the semiconductor layer, changing the resistance several orders of magnitude. Other manipulations, such as catalysis of an analyte in the solution or at the surface can change the chemical potential and also be transduced using this technique.

5.7 Self-aligned OFETs

Overlap capacitance can be a significant fraction of the total gate capacitance in a device. Overlap capacitance is especially problematic because it represents a load on performance without contributing to the transconductance of the device, and increases as a fraction of the total capacitance for a given design rule as the channel length decreases. This parasitic capacitance can be controlled by using high precision alignment and reducing the overlap margin allowed for source/drain alignment to the gate. An alternative approach is to deposit a high optical density gate and use this gate as a mask to align the source/drain layer. The process flow for a self-aligned process applied to the parylene encapsulation-based process described above is summarized in Fig. 5.12.

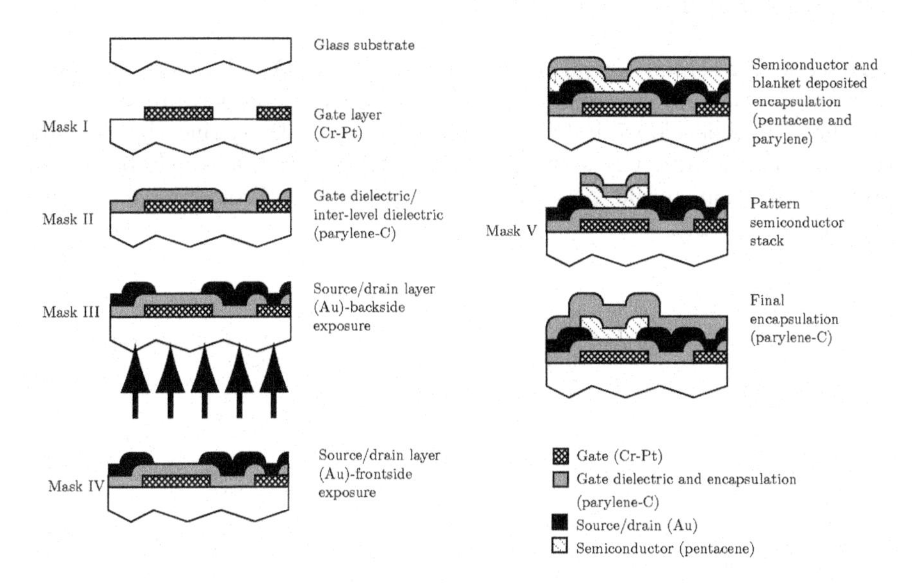

Fig. 5.12. A schematic showing the self-aligned process flow implemented in the parylene encapsulation process. An extra exposure step and mask is required to fill in interconnect shadowed by the gate layer, but these exposures are performed on the same photoresist and the total number of layers remains the same.

Using self-alignment, the source-gate and drain-gate overlap can be set to a minimum. Care should be taken to avoid leaving ungated material at the channel edge which can lead to significantly increased access resistance to the gated channel. The overlap/stagger gap distance can be managed by adjusting the dose delivered in the backside exposure step. The contrast edge will move toward the center of the channel as the dose increases.

The mask set required for this self-aligned process is slightly different than that required for the traditional front-exposure process. The source/drain layer must be patterned using a negative acting photoresist. The source/drain layer mask does not have a gap for the channel; this gap is provided by the gate layer. An additional front-side exposure source/drain level mask also needs to be included in the mask set to provide interconnect between the source/drain layer and the gate. This exposure is performed immediately before or after the backside exposure and before developing to fill in interconnect points which are otherwise shadowed by the gate layer. This step is necessary for most topologies; in addition to shadowing interconnect runs, the shadowing by the gate at via points can prevent reliable contacts from forming.

5.8 Conclusions

This section presented several advanced strategies for OFET processing which take advantage of improved knowledge of the structure/property relationships that have been developed in OFET material systems. By engineering the fabrication conditions, device structure, and materials used, it is possible to improve application specific performance and processing throughput in a range of material systems.

6

Modeling and characterization

6.1 Models

6.1.1 The role of models

Device models provide a common language for the discussion of the characteristics of transistor devices. The parameters measured in device characterization are the raw material with which the behavior of devices can be summarized using a few parameters and insights about the physical processes which underlie OFET behavior can be explained.

There are many excellent books and articles which discuss parameter extraction and device modeling of silicon and OFET devices in greater detail [112] [113]. This discussion will highlight the standard techniques for device characterization which have been used in OFETs, and also describe several of the newer techniques which are being developed which address the specific challenges that OFETs face.

6.1.2 The IEEE 1620 standard

The IEEE 1620-2004 standard covers the testing of OFET devices. This standard is actively under review and is periodically revised [114] and lays out a procedure for OFET parameter extraction which fits device curves to a simplified large signal long channel crystalline silicon (c-Si) device model with some adaptations for dealing with the complications OFETs present. While this approach has some limitations (see Section 6.5.1), to first order this approach will at least approximately reproduce this characteristic.

6.1.3 Long channel silicon device operation

Fundamentally, transistors are resistors where the charge carrier density is a function of a third terminal. This resistor model is the basis for all simple models of transistor behavior.

The IEEE 1620 method codifies the de facto standard practice of fitting OFET characteristics to a simplified long channel c-Si device characteristic which has its limitations, but is straightforward to implement. To simplify the discussion only the biases corresponding to NMOS c-Si devices will be presented. By convention, currents are positive when flowing into the device. Positive values of I_S, I_D, and I_G are the currents flowing into the source, drain, and gate, respectively. Voltages are traditionally referenced to the source, so V_{DS} and V_{GS} are the drain and gate voltages. L is the channel length (in the direction of current travel) and W is the device width (the dimension in the plane of current flow perpendicular to the length). C_{OX} is the specific gate dielectric capacitance in $F(cm)^{-2}$. Table 6.1 summarizes these parameters and the conventional (but mixed SI and cgs) units used in microelectronics.

Table 6.1. Traditional Si transistor parameters and conventional units for large signal modeling.

Parameter	Meaning	Units
I_S	Source current	A
I_G	Gate current	A
I_D	Drain current	A
V_{GS}	Gate voltage	V
V_{DS}	Drain voltage	V
W	Width	μm
L	Length	μm
C_{OX}	Gate capacitance (specific)	$F(cm)^{-2}$
μ	Mobility (effective)	$cm^2(Vs)^{-1}$
V_T	Threshold voltage	V
Q	Sheet charge density	C/cm^2

In traditional c-Si FETs, the source and drain are doped so as to create two back-to-back p-n junctions that block the flow of current in both directions in the channel. The gate is capacitively coupled to the channel across the gate dielectric. When the gate/channel capacitor is strongly biased to repel majority carriers and attract minority carriers, an inversion layer is formed at the channel/gate dielectric interface which is the same doping type as the diffused source and drain electrodes. This inverted layer short circuits the back-to-back diodes at the source and drain which previously held back the flow of charge, forming a resistor whose resistance is determined by the sheet charge density at the surface.

Inversion in a c-Si FET occurs when the gate voltage exceeds the threshold voltage (with the appropriate sign; in the positive direction of threshold in the case of an NFET). Right at threshold there is no charge in the channel, but the surface potential reaches a point at which it acquires a complementary character to its initial doping. Any further gate voltage applied beyond the threshold voltage accumulates more charge in the channel at a rate which is

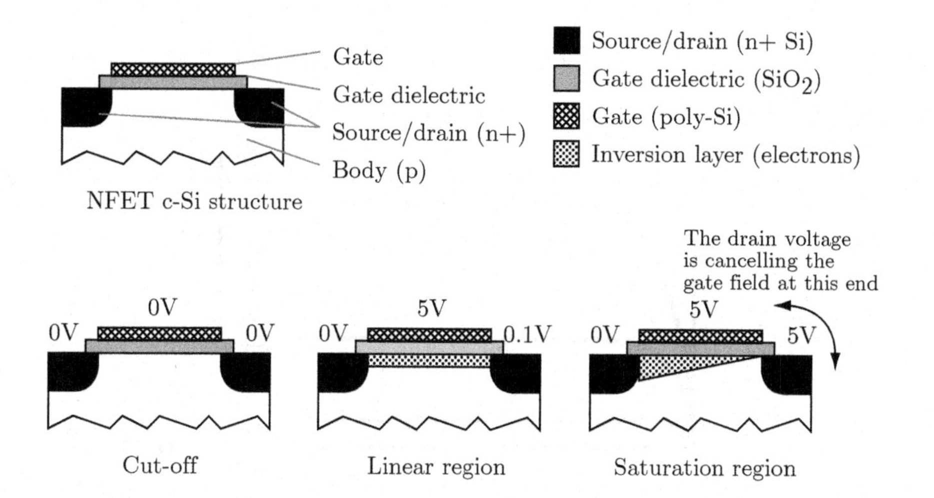

Fig. 6.1. A schematic showing the structure and regions of operation of a NMOS c-Si FET. The voltages shown are for illustration only. When V_{GS} exceeds the threshold voltage and V_{DS} is small the channel is inverted at both sides with approximately the same Q and the device behaves as a resistor. When a substantial V_{DS} is applied, the field induced by the gate is partially canceled on the drain end. When the potential at the drain end drops below V_T, $Q \approx 0$ and the carrier velocity increases to compensate, which leads to pinch-off and a saturation of the transistor characteristic. The carriers are all physically located very close to the gate dielectric interface, the triangle is illustrating that the carrier density is not constant. Since the current flow is constant across the length of the channel, the velocity and lateral field in saturation are not uniform.

determined by the capacitor relationship $Q_C = C_{OX}(V_{GS} - V_T)$ where C_{OX} is the capacitance of the gate dielectric and $(V_{GS} - V_T)$ is the gate voltage applied beyond the threshold voltage.

For a given gate voltage, as a larger V_{DS} (i.e. toward $+\infty$ for an NFET) is applied, the field generated by the drain voltage begins to cancel the field applied by the gate. This reduces the charge density at the drain end and reduces the rate of increase of current in the resistor formed by the accumulated channel as the V_{DS} is increased. When the drain end of the device has a potential equal to the threshold voltage (and, in this simple model, a free charge carrier concentration of zero), *pinch-off* is said to have occurred. The current stops increasing as V_{DS} increases, and the device then enters a region of operation known as saturation. The device is ideally seen as a current source with some second order increase in the current as V_{DS} changes due to an effective change in the channel length, but the c-Si model used for OFET characterization typically ignores this effect.

OFETs operate through a substantially different mechanism. The basic gate capacitor structure is the same (Fig. 6.3), but the contacts of OFETs are

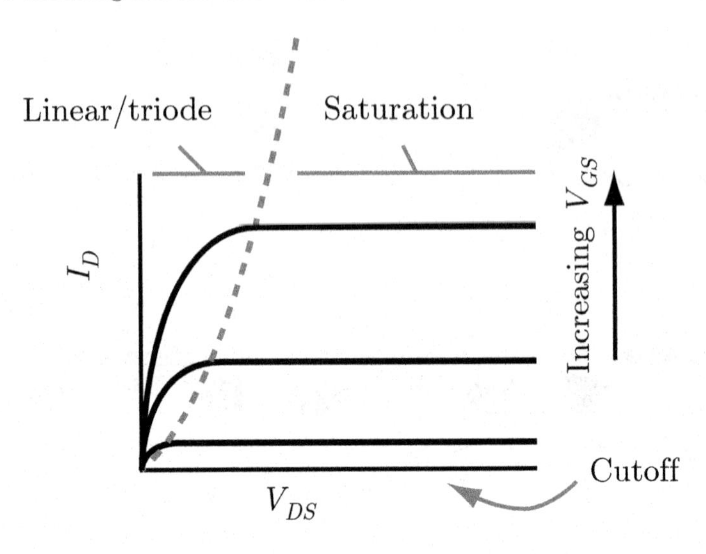

Fig. 6.2. A schematic diagram of the current flow in an FET with the regions of operation noted. The current in the saturation region can vary as a function of V_{DS} due to channel length modulation, but for the purposes of the simplified OFET model, where channels are very long, this effect is ignored.

engineered to have good electrical access to the channel. When the channel is depleted through the gate capacitor there are fewer mobile charges and less current flows. When the channel is accumulated, there are more charges and more conductivity. Just like in the c-Si transistor, as the V_{DS} is increased the channel current increases until the field from the drain cancels the gate field and applies less than the threshold field. At that point, I_D slows its rate of increase and the device appears to saturate.

From a modeling perspective, the threshold voltage is the gate voltage at which a perceptible current begins to flow in the device. For small voltages V_{DS}, where the gate of the transistor is biased beyond threshold, the device is a resistor whose resistance is determined by the gate voltage. For larger values of V_{DS}, the device is a current source whose magnitude is determined by the gate voltage. In a c-Si device below subthreshold, threshold can be defined where the drift current across the channel exceeds the diffusion current through the back-to-back diodes in the device (see Fig. 6.4).

6.1.4 Long channel silicon device model

To model its behavior to first order, the long channel silicon FET model separates the characteristic into three regions.

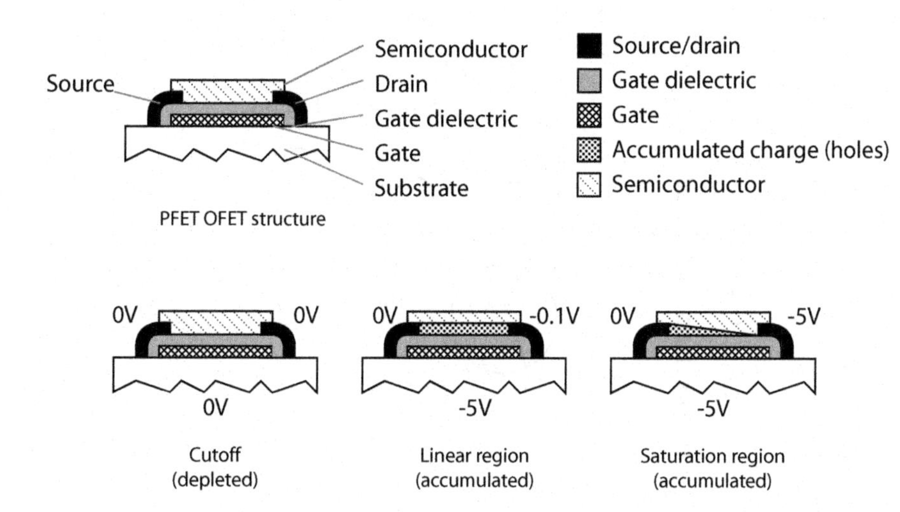

Fig. 6.3. A schematic showing the structure and regions of operation of a PMOS OFET. The voltages shown are for illustration only. The structure is inverted from the c-Si FET, but the operation is approximately the same. The major difference is that the channel current is only modulated by the density of charge in the channel and not by any barrier to injection from the source or drain electrodes.

Under biases where $V_{GS} < V_T$, we are in cutoff and no current flows (in a silicon FET, because of the back-to-back pn diodes, in an OFET because of no available mobile carriers).

$$I_D = 0A \tag{6.1}$$

At small V_{DS} and larger V_G, where $V_{GS} - V_{DS} > V_t$, the transistor is said to be in the linear region. This is modeled as a resistor whose resistivity is determined by the mobility and geometrical factors ($\mu \frac{W}{L}$), and the charge carrier density is determined by the capacitance and the applied voltage beyond the threshold ($C_{OX} * (V_{GS} - V_t)$).

This gives a current in the linear region (where $V_{GS} > V_T$ and $V_{DS} < V_{GS} - V_T$) of:

$$I_D = \frac{W}{L} \mu C_{OX} (V_{GS} - V_t - \frac{V_{DS}}{2}) V_{DS} \tag{6.2}$$

As V_{DS} is increased, when $V_{GS} - V_{DS} < V_T$, the current reduces its rate of increase because of the onset of pinch-off. We can determine the current at the saturation point by substituting $V_{DS} = V_{GS} - V_T$ in the linear region equation and simplifying to determine:

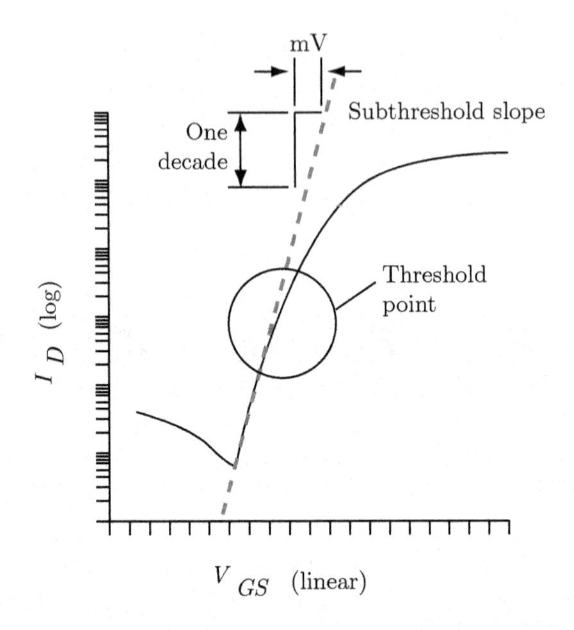

Fig. 6.4. A schematic showing the subthreshold voltage/current characteristic and the subthreshold slope extraction. The threshold voltage can be taken as the point at which the current deviates from the exponential character expected from purely diffusive transport.

$$I_D = \frac{1}{2}\frac{W}{L}\mu C_{OX}(V_{GS} - V_t)^2 \tag{6.3}$$

This simplified model does not capture the behavior in deep subthreshold, changes in current in the saturation region, gate leakage, and parasitic series resistances. These parameters will be considered separately and need to be included in comprehensive device models.

6.2 Parameters

6.2.1 Mobility

Fundamentals of mobility

Mobility is the relationship between the carrier speed in a material and the applied electric field. Microscopically, carriers accelerate due to the force applied by the electric field and periodically scatter and experience other momentum transfer phenomena which dissipate the acquired momentum. This occurs rapidly and across a large number of carriers which averages the observed effect. Observed macroscopically, carrier velocity is a viscous flow process in solid materials. The velocity appears constant for a given electric field and is linearly dependent on the field. Mobility is defined by:

$$\text{Velocity} = \mu E \tag{6.4}$$

$$\mu = \frac{\text{Velocity}}{E} \tag{6.5}$$

And as a consequence the units (in the conventional microelectronic unit system) are in $\text{cm}^2(\text{Vs})^{-1}$.

The current which flows is the amount of charge contained in a volume swept at the speed of the charge carrier movement. For a sheet charge density Q ($\text{C}(\text{cm})^{-2}$) and a width W, the amount of charge contained in a solid swept out in this manner yields a current of:

$$I = \frac{\text{Charge}}{\text{Unit time}} \tag{6.6}$$

$$= QWx \tag{6.7}$$

$$= QW\mu E \tag{6.8}$$

And therefore the mobility is defined by:

$$\mu = \frac{I}{QWE} \tag{6.9}$$

This model assumes a linear relationship between carrier velocity and electric field, which is generally true at low electric fields. In crystalline semiconductors, the relationship between velocity and electric field is nonlinear at high fields. In organic semiconductors, the relationship is non-linear at larger electric fields and also dependent on the charge carrier concentration. This makes the interpretation of mobility somewhat more complicated than fitting to a linear model.

It is possible to measure the mobility of carriers directly in the linear region of operation [115]. This is an alternative to determining the mobility through curve-fitting to a transistor model. The mobility measured this way has an unambiguous meaning, but the approach has some limitations. The total current in the transistor deep in the linear region can be treated like a resistor:

$$\mu = \frac{\text{Velocity}}{E} \tag{6.10}$$

$$Q = \frac{C_{channel}(V_{GS} - V_T)}{WL} \tag{6.11}$$

$$\mu = \frac{I_D}{WQ}\frac{L}{V_{DS}} \tag{6.12}$$

$$= \frac{I_D}{W}\frac{WL}{C_{channel}(V_{GS}-V_T)}\frac{L}{V_{DS}} \tag{6.13}$$

$$= \frac{I_D L^2}{C_{channel}(V_{GS}-V_T)V_{DS}} \tag{6.14}$$

While not all of the charges which are measured in capacitance are mobile, this model has the advantage of not relying on a transistor model for its operation, it is based only on the fundamental definition of mobility. The threshold voltage and channel charge can be extracted from measurements of the gate capacitance (see Section 6.2.2). This method is only effective in the linear region where the charge carrier density is approximately constant across the channel, i.e. where V_{DS} is small (<0.1V or less). Because charge is conserved in the channel:

$$I_D = Q(x)E_{lateral}(x)\mu(x) \tag{6.15}$$

where $x$is the position in the charge carrier flow direction, $E_{lateral}$ is the local electric field in the charge carrier flow direction, and I_{DS} is constant as a function of x due to conservation of charge.

For small values of V_{DS}, $Q(x)$ is approximately constant because the field across the V_{GS} and V_{GD} interfaces are similar. The carrier velocity, $Elateral(x)\mu(x)$, is therefore constant. As the device begins to pinch off, Q becomes a stronger function of x, and the carrier velocity and lateral field are no longer constant across the length of the channel. In this case this method of mobility extraction is not valid.

6.2.2 Threshold voltage

The measurement of threshold voltage is somewhat complicated by the relatively gradual turn-on of the device. In OFETs (as in c-Si devices), there is no single definition which is universally accepted.

Properly speaking the threshold voltage in c-Si is the voltage at which inversion is achieved in a metal-insulator-semiconductor (MIS or MOS) capacitor. Because organic semiconductors never achieve inversion, strictly speaking there is no definite threshold voltage, only a voltage at which the device begins accumulating charge (which, in c-Si devices, would be referred to as the flat-band voltage). Despite this difference, the term threshold voltage is generally used in the OFET community to specify the gate voltage at which current begins to flow which in c-Si FETs occurs with the onset of inversion, and in OFETs at the onset of accumulation.

One approach was shown in Fig. 6.4. The threshold voltage is reached when the current increase deviates from the characteristic expected from diffusion in a c-Si device. This point is usually at a small current, but shows where the influence of the gate starts to overcome the barrier to conduction in the channel.

Another simple method for determining the threshold voltage is to extrapolate the linear part of a $\sqrt{I_D}$-V_{GS} curve to the intercept on the V_{GS} axis (see Fig. 6.5). This, to first order, indicates the voltage at which there is some increase in current and the device transitions from cut-off to a region of operation in which the device is conducting current. This method is convenient, simple to implement, and provides a definition of threshold which is understandable from a current flow perspective (i.e. when current starts to flow, V_{GS} has exceeded V_T).

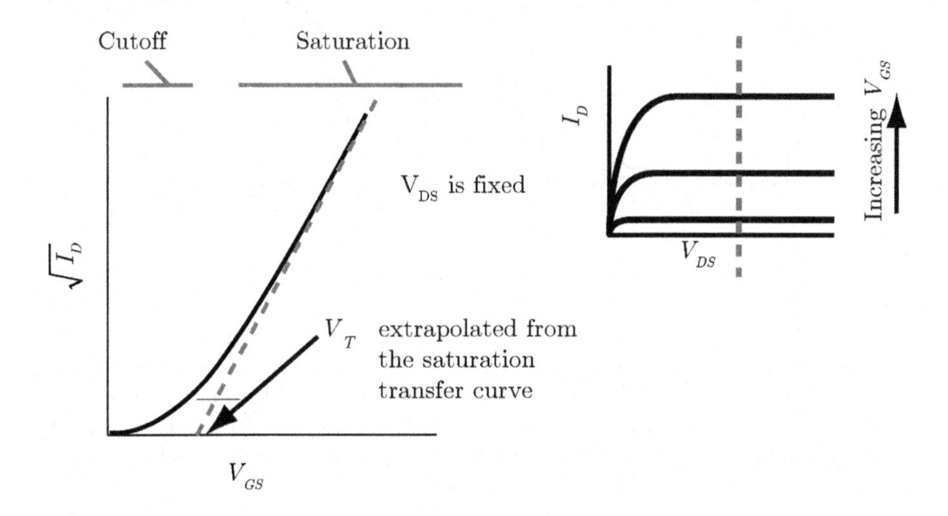

Fig. 6.5. A figure schematically showing the extrapolation of V_T from the $\sqrt{I_D}$-V_{GS} transfer characteristic. The inset schematically shows the section of the $I_D - V_{GS}$ output characteristic where the transfer curves are taken.

There are several limitations to the extrapolation approach:

- The turn-on in the subthreshold region of operation is not as abrupt as in a high quality crystalline semiconductor device. Significant current can flow before the device enters the extrapolated higher transconductance region in the transfer curve.
- The part of the linear and saturation which one may expect to be linear (based on the model) may not be; this leads to some ambiguity in the threshold voltage determination

- The extrapolated zero point has no universally agreed upon physical significance; it is the extrapolation of the expression for $\sqrt{I_D}$ into a region in which the model does not apply

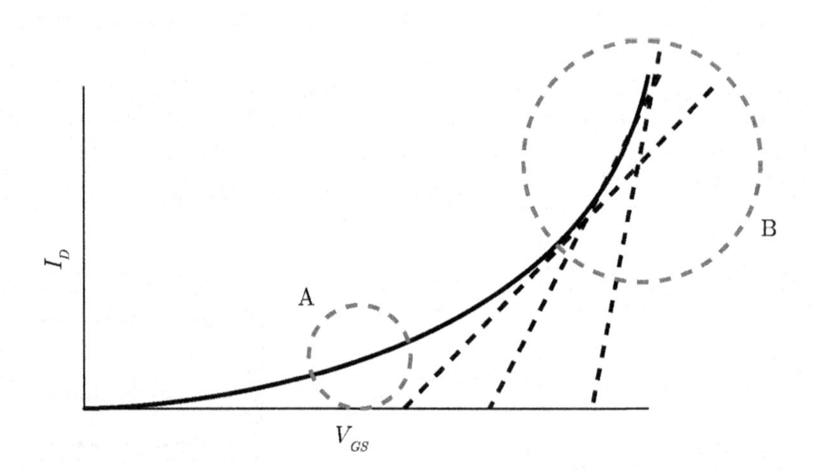

Fig. 6.6. A figure showing some of the limitations of determining the threshold voltage through linear extrapolation. In region A, there is significant current flow below the nominal V_T. This is a consequence of slow subthreshold turn-on in the device. Region B shows a non-linear current increase in the triode region of device operation. This type of increase (which can, for example, be caused by field dependent mobility) causes ambiguity in the extrapolated threshold value.

A third strategy for determining the threshold voltage is to measure the capacitance-voltage characteristic of the transistor (or $C-V$ curve), which directly indicates the carrier density [116]. The capacitance is measured between the source and drain held together and the gate (Fig. 6.7 (a)).

In accumulation (Fig. 6.7 (b)) there are three contributions to capacitance: the channel capacitance $C_{OX}WL$, the gate-source overlap capacitance C_{GS}, and the gate-drain overlap capacitance C_{GD}. As the gate bias is adjusted into a region where the device is depleted, (Fig. 6.7c) the observed capacitance is only the geometrical capacitance of the source/drain overlap and fringing fields $C_{GD} + C_{GS}$. There are no mobile carriers in the channel to contribute to a channel capacitance with respect to the gate.

The point in the gate voltage sweep at which mobile charges in the channel appear or disappear can be relatively clearly observed this way, and the voltage at which accumulation in the channel begins can be interpreted as the threshold (or flatband) voltage.

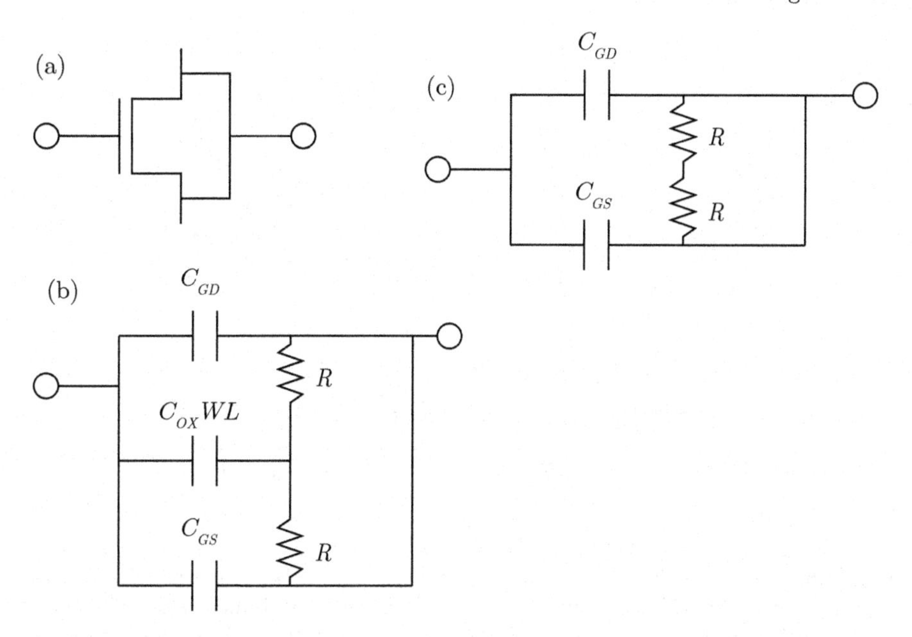

Fig. 6.7. Sschematic models for $C - V$ measurement of OFETs. (a) shows the connection made to take the characteristic; V_{DS} is set to 0V and the capacitance is measured between the gate and the shorted source and drain electrodes. In accumulation (b) there are three major contributions to the total capacitance; the channel capacitance $C_{OX}WL$, the gate/source overlap capacitance C_{GS}, and the gate/drain overlap capacitance C_{GD}. In depletion (c), only the overlap capacitance is observed, there is no mobile charge in the channel to contribute to a channel capacitance. The transistor channel is really a distributed RC structure and the lumped R shown is only schematic.

Because charges need to move in and out of the channel to observe the channel capacitance, the RC effective cutoff of the channel is of importance. Fig. 6.7 (b) lumps the channel resistance into $2R$, but the channel is a distributed RC structure which charge must pass through to participate in the capacitance measurement. The R-C cutoff of OFETs can be quite low (sometimes kHz or lower), and depending on the observed device performance, the use of a low frequency or (ideally) quasi-static DC capacitance measurement is essential. This allows adequate time for charges to move in and out of the channel.

This method has several advantages and disadvantages over the use of I-V curve fitting to determine the flatband voltage:

Pros:

- Contact resistance and other resistive parasitics do not factor into the threshold voltage measurement if it is performed in DC

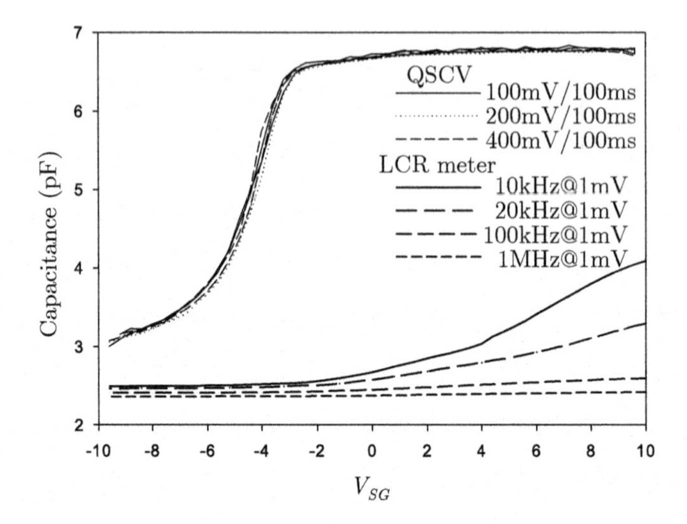

Fig. 6.8. A CV curves from a pentacene OFET with a W/L of $1000/20\mu m$ from [117]. The quasi-static CV captures the entire device capacitance, whereas the capacitance bridge measures too quickly to fully charge the channel and reads an incorrect value.

- Mobile carriers appear relatively abruptly, which makes the measurement less ambiguous
- The total charge in the channel can be determined with this measurement by integrating the area under the graph

Cons:

- Mobile charges appear in capacitance measurements before current begins to flow; little current flows right at the threshold voltage which makes this value disagree with transconductance-based models
- The measurement cannot easily be adapted to higher frequency device measurements because of parasitic limitations

6.2.3 Contact resistance

Contact resistance (R_C) includes all series resistance that does not scale with channel length. Contact resistance will include resistance attributable to the source and drain composition, the interface between the semiconductor and the contacts, and edge induced morphological or other structural changes which change the channel resistance but do not operate in the center of the channel.

Measuring contact resistance is relatively straightforward. A model for contact resistive effects is shown in Fig. 6.9.

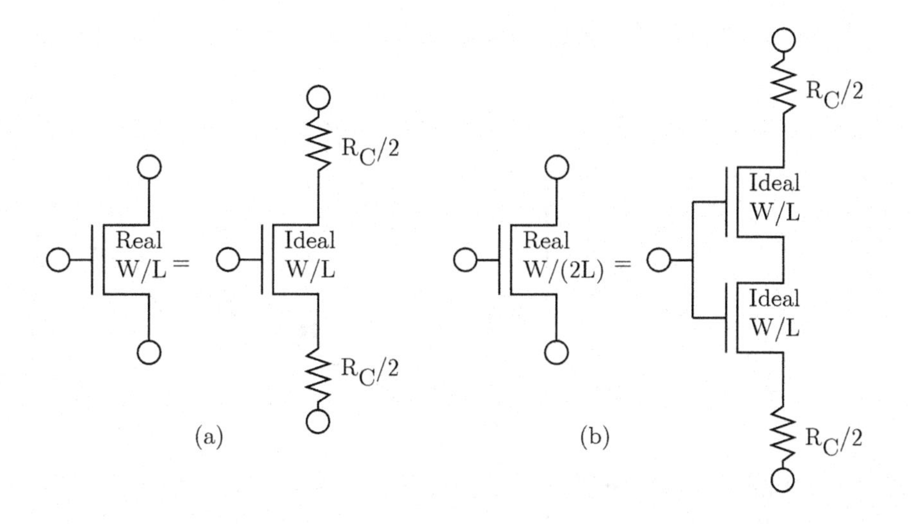

Fig. 6.9. A model for contact resistance. A real device is modeled as an ideal device which exhibits no contact resistance and a contact resistance in series at the source and drain (a). If a longer but otherwise identical device is fabricated, the contact resistance should remain constant, while the ideal transistor scales (b). Comparison of different length devices is the basis for the transfer line method of transistor contact resistance extraction.

Mathematically, if R_C is taken to be the total resistance attributable to the contacts (in ohms) and $\rho_{channel}$ is the specific channel resistance at the bias condition in question (in ohms/micron) the total device resistance can be given as:

$$R_{total} = R_C + L * \rho_{channel} \tag{6.16}$$

where L is the channel length (i.e. the distance that carriers travel between source and drain). By comparing the channel resistance in a series of transistors of different lengths under comparable bias conditions, the $L = 0\mu m$ intercept can be easily determined and the contact resistance extracted. If the contact resistance is a linear resistance, the magnitude of the lateral electric field $\frac{V_{DS}}{L}$ is not significant, but to minimize the influence of non-linearities on the contact resistance measurement it is generally advisable to scale V_{DS} in the measured points to keep $\frac{V_{DS}}{L}$ constant. If the intersection point is not at $L = 0$, there is a systematic variation which has led to an offset in the channel length (e.g. undercutting of the electrodes or mask over/under exposure).

For a given process and bias the contact resistance scales with the inverse width of the transistor. This specific contact resistance is specified by some authors in ohms/micron (Ω/μm).

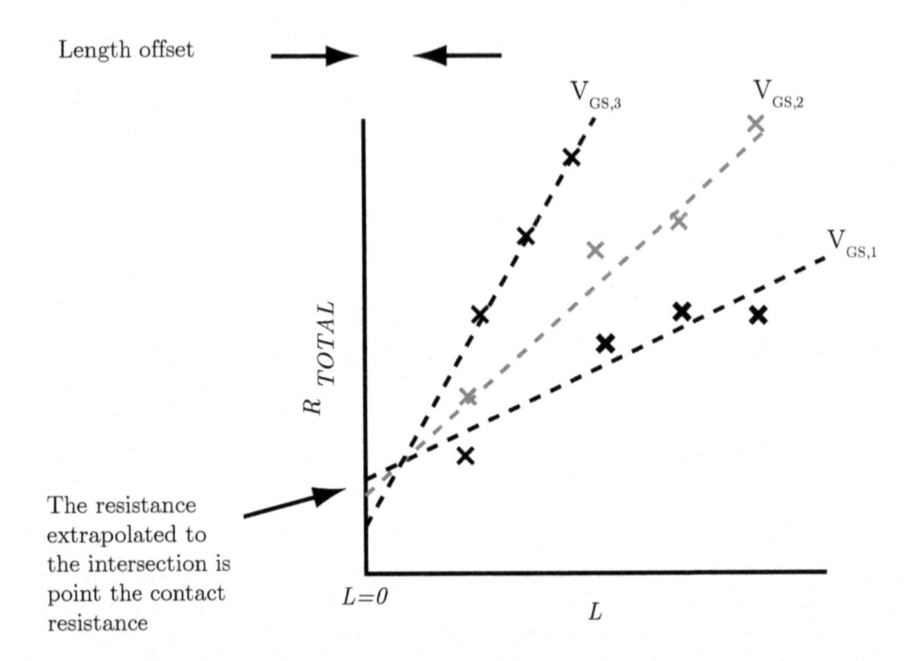

Fig. 6.10. A schematic demonstrating extraction of the contact resistance from measurements of different length devices. The total channel resistance for a set of devices of different lengths are plotted, where $R_{total} = R_C + L * \rho_{channel}$. At the intersection point (ideally $L = 0$, but a systematic offset in the length can cause this point to be elsewhere), $R_{total} = R_C$
. Different gate biases can also be plotted and the corresponding contact resistance measured.

This technique can be applied at a range of gate biases and plotted as a function of channel charge density, gate voltage, or applied lateral electric field. Non-linear models have been proposed for the contact resistance as a function of the applied bias (see, for example, [118]), which generally assume a Schottky or other diode-like charge injection from the contact into the channel. The transfer line method can also be used to extract a purely empirical model.

Contact resistance in OFETs can be a substantial fraction of the total channel resistance. Because its magnitude depends on the applied bias, separation of the contact resistance from other device measurements (e.g. mobility) has been shown to significantly improve the analysis of device performance data [115] [119].

6.2.4 Hysteresis/bias-stress

Hysteresis in drive characteristics is problematic. Hysteretic responses are, by their nature, challenging to model since the transistor retains a history of drive and the response depends on that history. Hysteretic device characteristics also combine the transport of carriers, which are induced by the applied bias, and carriers which are released by traps or dielectric relaxation, which can lead to errors in parameter extraction.

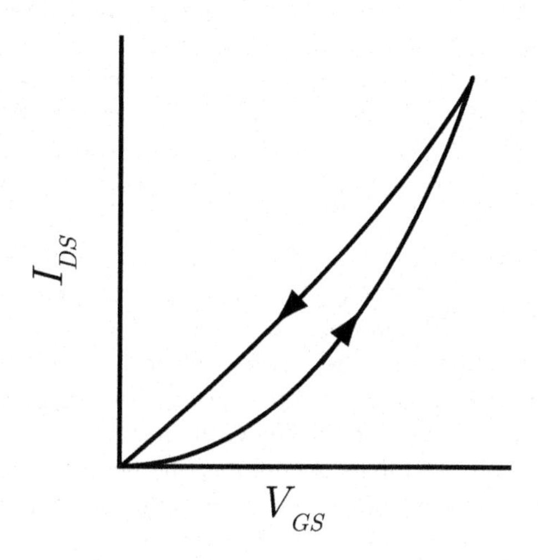

Fig. 6.11. A schematic transfer characteristic exhibiting hysteresis.

While it is generally preferable to have devices with as little hysteresis as possible (with the exception of devices designed to store state), there are a few precautions that can be taken to monitor the effect of hysteresis:

- Questionable *I*-*V* characteristics should be measured in both directions (double sweeps) at the same sampling speed; this serves as a qualitative measure of bias-history hysteresis,
- Hysteresis can be measured using sweeps which are at the frequency of interest. DC hysteresis may be immaterial for AC circuits.
- Device measurement can be performed after the device is held at a fixed bias for some time; this will expose the direction and nature of the hysteresis. The IEEE 1620 standard recommends 10 minutes at $V_{GS} = V_{DS} = 0V$. Holding the device at a high bias before measurement can also be used to measure the complementary effect.

Several studies have identified three major contributions to short term hysteretic effects [94]:

- Slow relaxation of the gate dielectric
- Dielectric charge storage

- Traps in the semiconductor

Slow relaxation of the gate dielectric is usually strongest in polymer-based gate dielectrics which have polar groups [120] or absorb water [121]. The relaxation of the gate dielectric invariably increases the dipole in the gate dielectric as time passes. This causes a decrease in current after a bias change which increases the current flow.

Dielectric charge storage is caused by injection of charges into the gate dielectric. The driving force for the relaxation increases the field drop across the gate dielectric and decreases the field which reaches the semiconductor. This generally leads to a decrease in current after a bias change which increases the current flow.

Semiconductor traps have been directly observed in transistors and organic semiconductor crystals by a variety of techniques including temperature dependent current analysis [122], photoluminescence [123], and transient current spectroscopy [70], among others. These traps can increase or decrease the observed current depending on their charge; if they trap majority carriers the current will decrease over time, but if they emit majority carriers (and therefore hold a charge which reinforces accumulation) current can increase over time. These traps are often extrinsic (water and oxygen often form these centers [124]) and their formation is sometimes reversible. Understanding their nature can lead to insight on how to process and encapsulate devices to prevent their formation.

In addition to reversible hysteretic effects, irreversible storage, photoinduced, and bias-stress can also be observed in many OFET systems. Repeated characterization through the application of the stressor will expose these effects.

6.2.5 Gate leakage

Ideally $I_G = 0\text{A}$ at all bias conditions. Imperfect gate dielectrics, surface conduction, bulk device transport, or a lack of semiconductor patterning can lead to gate leakage and affect the performance of many circuits.

Conservation of charge requires that:

$$I_D + I_S + I_G = 0 \tag{6.17}$$

In an ideal FET with no gate leakage ($I_G = 0$) and $I_D = -I_S$.

The gate current can be measured while other I-V measurements are taken. The simplest measure of leakage is taken while $V_{DS} = 0$ over a range of V_{GS}, but the gate current can be measured over a range of bias conditions and accounted for in circuits where it can have an effect.

6.2.6 Subthreshold slope

The subthreshold slope is the inverse slope of the logI_D vs V_{GS} measured below threshold. It is typically reported in base 10 logarithmic units of mV/decade (i.e. mV of gate bias for decade of drain current modulation). Smaller values in these units correspond to a larger slope, which is generally more desirable. Fig. 6.4 shows the extraction of subthreshold slope.

Traditionally, subthreshold slope is reported at the area of maximum slope. In many thin film transistors the current begins to rise again at bias points which are normally associated with cut-off. This rising current is generally because of increased gate leakage ($|V_{GS}|$ and $|V_{GD}|$ continue to increase). This may alternatively correspond to ambipolar behavior and a second region of accumulation in some materials [125] or hole and electron transporting semiconductor blends [126].

Current in the channel in subthreshold is a function of charge carrier concentration in the channel. The best subthreshold slope which can be observed in an FET device at room temperature is 60mV/decade, which is the slope at the edge of a Fermi distribution when it is convolved with an abrupt density of states. Because organic semiconductors exhibit a gradual rise in the density of states at the channel edge and not all carriers are equally mobile, the convolution produces a shallower rise in the carrier density and the observed subthreshold slope is worse than this value.

Measurement of the subthreshold slope is simple; for a V_{DS} of interest, the slope can be taken at the steepest point of the log-linear I_D vs V_{GS} curve.

Subthreshold slope is a parameter in which hysteretic effects can complicate data analysis; gate dielectric polarization relaxation and charge trapping/detrapping can apply an additional potential swing to a device beyond that which is applied through the gate bias. This can lead to the measurement of larger or smaller values of the subthreshold slope; measurement of the slope in both increasing and decreasing gate voltages at the same sampling speed can diagnose this problem and indicate if a slower measurement speed is warranted.

6.2.7 Output conductance

The device characteristic in saturation does not exhibit a perfect current source characteristic as V_{DS} is increased; the current generally continues to increase with some rate. In silicon transistors this is to first order a manifestation of channel length modulation. While channel length modulation is theoretically possible in OFETs with a small L, parasitic parallel conduction paths are the primary contribute to the output resistance of OFETs in saturation.

The output conductance and resistance are calculated as:

$$g_o = \frac{\Delta I_D}{\Delta V_{GS}} \tag{6.18}$$

$$r_o = \frac{\Delta V_{GS}}{\Delta I_D} \tag{6.19}$$

This can be easily extracted from the slope of the gate sweep taken at the bias point of interest.

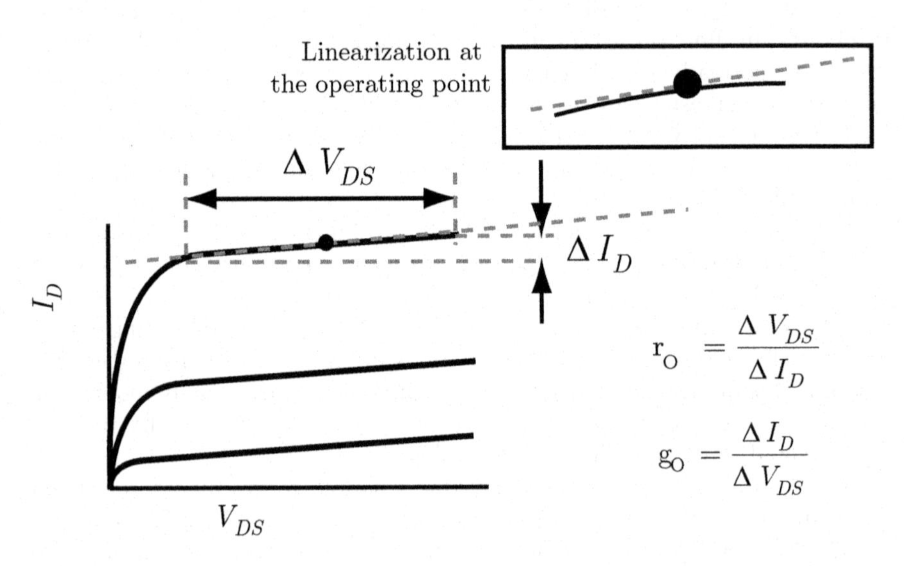

Fig. 6.12. Output conductance and resistance as observed in the drain sweep/output characteristic. The inset shows the linearization of the output characteristic used to determine the conductance at the operating point.

6.3 Characterization

The selection of tests used to characterize a device needs to be motivated by the end application of the transistors in question. There are, however, some general guidelines which can guide the structure of test programs. This section makes some recommendations and also includes information on the recommendations of the IEEE 1620 standard.

Device current-voltage measurements are typically taken using a semiconductor parameter analyzer under automatic control operating in near-DC. The same measurements can also be taken at higher frequencies using a function generator, high speed parameter analyzer, or curve tracer to determine these parameters at higher frequencies. Ideally all three terminals' currents and voltages should be measured during analysis. Capacitance measurements are necessary for complete device characterization. These measurements should be performed using either a low frequency capacitance bridge or more ideally a system configured for quasi-static capacitance analysis.

Choice of a test vehicle is important. It is best to use devices which are fabricated in the same process that will be used to make circuits based on the model. A convenient choice is to use oxidized doped Si wafers as the gate and gate dielectric. While this structure has some advantages for scientific studies, it is not a physically realistic vehicle in which to test device performance and carries some hazards. Si does not form a true metallic capacitor; the structure measured is a function of both the parasitic MOS capacitor formed by the substrate and the transistor behavior in the organic semiconductor. Care should be taken to select a substrate doping which has threshold and flatband occur outside bias levels of interest for the relevant measurement. The gate-source and gate-drain capacitance is also necessarily large in such a structure because of the unpatterned gate. This complicates the measurement and interpretation of capacitance dependent data and slows the device operation in dynamic measurements.

6.3.1 Gate sweep/transfer characteristic

Basic transistor action is measured by holding V_{DS} constant at a few values and sweeping the gate voltage. At least one value of V_{DS} should be selected to test the device in the linear region and at least one in the saturation region. For pentacene transistors which are, for example, designed to operate at -20V, these curves could at a minimum be taken at $V_{DS} = -0.1$V for the linear region measurement and $V_{DS} = -20$V for saturation. Gate voltages could be swept over a range of $V_{GS} = 0$ to $-(-20)$V, or perhaps a little further.

The linear region should be plotted as I_D vs. V_{GS}. The saturation region characteristic can be plotted as $\sqrt{I_D}$ vs. V_{GS} or on a linear curve. These curves are often plotted together in the literature using split axes to conserve space. Velocity-based charge carrier mobility can be determined from $V_{DS} = 0.1$V when the total charge from the capacitance measurements or charge modeling are known. The determination of threshold voltage from extrapolation from these graphs can also be performed.

The linear region measurement should ideally be taken at as low a value of V_{DS} as possible. It is only for low values of V_{DS} that the electric field and charge density in the channel is uniform and the measurement of mobility is valid. Mobility varies significantly as a function of gate voltage, it should be measured and reported as a function of V_{GS} by taking the slope tangent at each V_{GS} individually.

6.3.2 Drain sweep/output characteristic

The small signal device characteristics in saturation can be determined from an additional sweep taken at a finite number of values of V_{GS} over a more

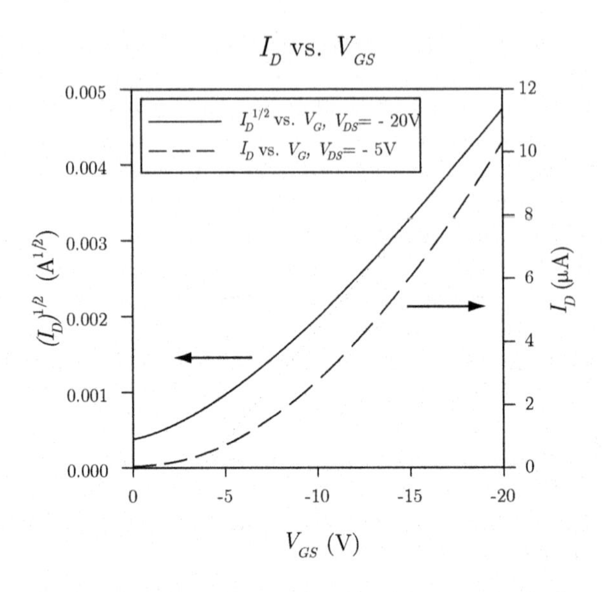

Fig. 6.13. A sample device transfer measurement showing both the saturation and linear characteristics on one graph.

finely divided range of V_{DS}. For the hypothetical pentacene transistor, for example, this range could be $V_{GS} = 0, 4, 8, 12, 16$ and 20V; and $V_{DS} = 0$ to -20V. The output conductance can be determined from this graph.

The output characteristic is almost universally presented as the qualitative proof that a transistor has been formed.

6.3.3 Capacitance

The capacitance of the transistor is ideally measured using a quasi-static CV method which uses either integration of charge from voltage steps or measurement of current in a ramp. A low frequency capacitance bridge can also be used; the maximum usable frequency depends on the transistor mobility and total parasitic capacitance in the test structure. Because charges move slowly in low mobility materials and the parasitic capacitances in thin film transistors can be substantial, the RC cutoff of the transistor can complicate the measurement of capacitance using a capacitance bridge. To effectively use quasi-static methods, however, the gate leakage must be small relative to the current seen on each step.

Capacitance can be measured by setting $V_{DS} = 0$ and sweeping V_{GS} over the range of interest. The capacitance measured while the device is in depletion is the overlap capacitance $C_{GS}+C_{GD}$ (where C_{GS} and C_{GD} are the total overlap capacitance in F). Once the bias passes flatband and the device accumulates, the capacitance increases to $C_{GS} + C_{GD} + C_{OX}WL$ (where C_{OX}

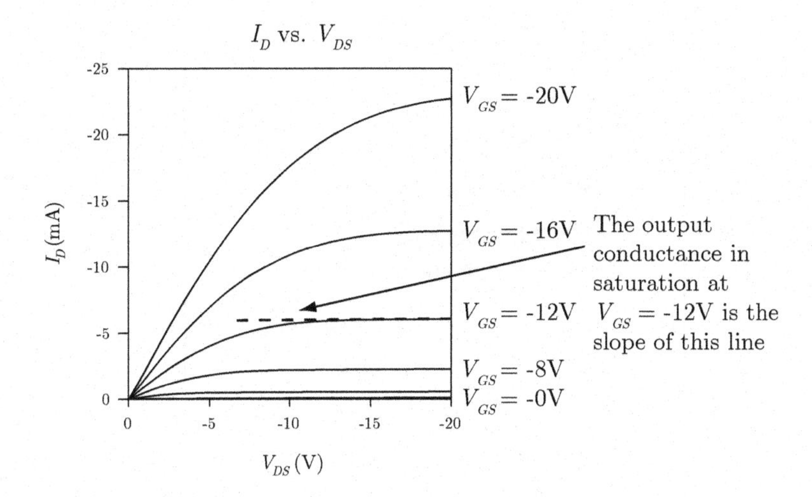

Fig. 6.14. The output characteristic for a pentacene OFET. The output conductance can be determined from this curve at each gate bias condition.

is the specific gate dielectric capacitance, in $F(cm)^{-2}$. The electrostatics of OFETs are somewhat different from c-Si FETs. There is no body; the gate sees capacitance with only the electrodes and the channel when it is accumulated. While there is also a contribution due to fringing capacitance, this is generally negligible in thin film transistors which are not self-aligned. Electrostatic modeling can be used to determine the fractional contribution to overlap and fringing. In most transistor processes it is also convenient to make a test metal-insulator-metal capacitor between the source/drain layer and gate layer in order to directly measure the gate dielectric capacitance. When reporting a structure with a well characterized gate material, the gate capacitance can also be specified as a relative dielectric constant (K) and thickness (t_{ox}).

6.3.4 Gate leakage

Gate leakage can be measured by adding a gate sweep to the measurement program with $V_{DS} = 0V$. No explicit figure of merit has been established for gate leakage (which is generally non-linear), but I_G can be plotted as a function of gate bias.

6.4 Device model

The complete large signal and small signal static model is provided by Equations 6.2, 6.3, and 6.1. For dynamic modeling, a simple small signal model

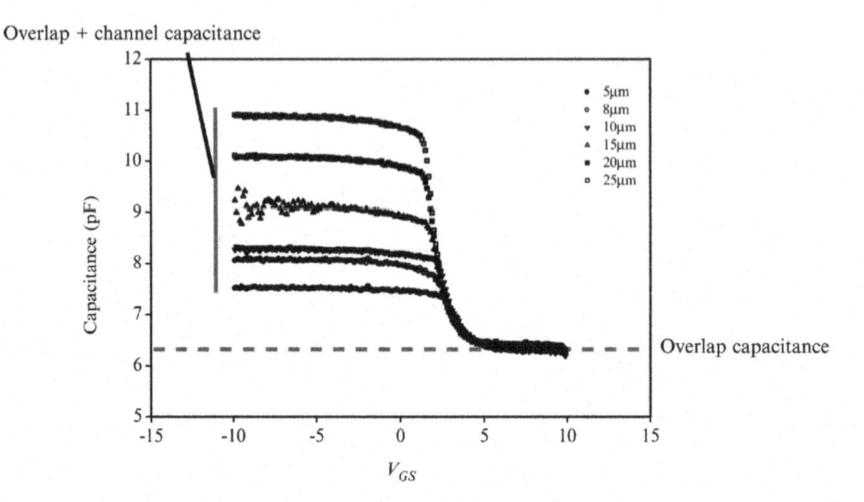

Fig. 6.15. C-V characteristics taken using quasi-static methods for a series of devices with the same source and drain overlap, but different channel lengths. In depletion (toward the right part of the graph) the capacitance observed is only the overlap capacitance $C_{GS} + C_{GD}$ and is the same for all devices. In accumulation the capacitance is $C_{GS} + C_{GD} + C_{OX}WL$, and the channel capacitance can be measured from the difference.

based on the linearized parameters discussed in this chapter is given in Fig. 6.16. The transconductance, g_m, is determined from the transfer characteristic and is defined as:

$$g_m = \frac{\partial I_D}{V_{GS}} \tag{6.20}$$

Any parallel series resistance (e.g. from too thick a semiconducting layer) can be encapsulated in r_o. The contact resistance is considered externally from the basic device and should ideally be considered separately from g_m and r_o. If it is not characterized separately, it will instead lumped with g_m and r_o at the relevant bias point.

6.5 Parameter summary

At a minimum the 1620 standard requires reporting of the mobility, threshold voltage, and on/off ratio of the device. It also requires reporting of the measurement conditions (temperature and relative humidity) as well as the W, L, and enough information to determine the gate capacitance. In the literature mobility is usually measured by curve fitting in the saturation region, which

Table 6.2. OFET parameters, the corresponding required measurements, and status with respect to the 1620 standard. The 1620 standard specifies that the mobility should also be extracted in the saturation region via curve fitting, but this measurement is suspect, see [115] and Section 6.5.1.

Parameter			Required measurement			
Name	Symbol	Units	Gate Sweep	Drain Sweep	Capacitance	1620 Standard
Gate capacitance (specific)	C_{OX}	$F(cm)^{-2}$			X	Required
Gate dielectric loss tangent	d	degrees			X	Optional
Linear region mobility (transconductance-based)	μ	$cm^2(Vs)^{-1}$	X			Required
Linear region mobility (velocity based)	μ	$cm^2(Vs)^{-1}$	X			
Saturation region mobility*	μ	$cm^2(Vs)^{-1}$	X			Required
Threshold voltage (extrapolation based)	V_T	V	X			Required
Threshold voltage (capacitance based)	V_T	V			X	
Gate leakage	I_G	A	X			Optional
Source/drain capacitance	C_{GS}	F			X	Optional
On/off ratio	I_{ON}/I_{OFF}	Unitless	X			Required
Subthreshold slope		V/decade	X			Optional
Channel charge	Q	$C(cm)^{-2}$			X	
Contact resistance	R_C	Ω/cm	X			
Output resistance	r_o	S		X		
Transconductance	g_m	S	X			

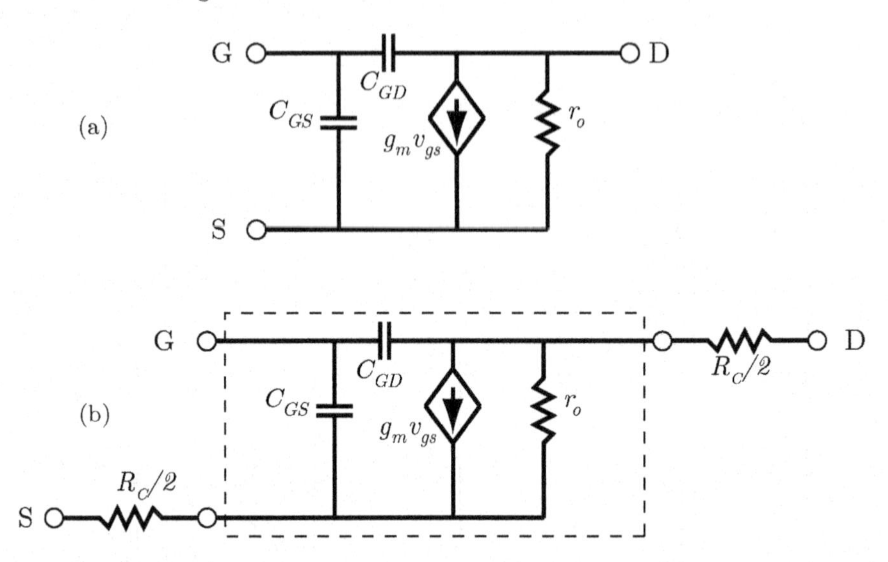

Fig. 6.16. A simple small signal model for OFET devices. (a) shows the model of the core device, (b) shows the device with contact resistance characterized and included in the model.

in typical devices yields a value of 0.1-1 cm^2/Vs, and the on/off ratio is measured from maximum to minimum current value and is usually gate leakage or channel leakage limited to $10^4 - 10^6$.

6.5.1 The limits of curve fitting in amorphous systems

Curve fitting to c-Si device models poses several risks, because the assumptions inherent in the model are not necessarily valid in disordered semiconductor systems. Of particular note are the lack of a single uniquely definable mobility and the lack of a well defined threshold voltage. Both of these characteristics lead to inaccuracies in modeling which have led to the adoption of other transport models based on amorphous silicon (a-Si) or polysilicon (p-Si) device models.

Both the lack of a sharp definable mobility and threshold voltage are attributable to a gradual density of states profile at the valence or conduction band (HOMO/LUMO) frontiers of the material. Because carriers are relatively localized and hop from molecule to molecule, a continuum of states means that there is a distribution of energy barriers to conduction. As the Fermi level of the semiconductor moves closer to a higher density of states, states closer to the edge (which appear more mobile) are populated, and the incrementally added new carriers are significantly more mobile than deeper charges which were added earlier [127].

Table 6.3. Device performance, structure, and test parameter table for IEEE 1620 compliance, adapted from [114], Fig. 4. At least two of the specific gate insulator capacitance, relative dielectric constant, and gate dielectric thickness are required. Reprinted with permission from IEEE Standard 1620-2004, IEEE Standard for Test Methods for the Characterization of Organic Transistors and Materials, Copyright 2004 by IEEE.

Parameter	Symbol	Units
Channel width	W	μm
Channel length	L	μm
Temperature (at measurement)	T	°C or K
Relative humidity (at measurement)	RH	%
Mobility	μ	$cm^2(Vs)^{-1}$
Threshold voltage	V_T	V
On/off ratio	I_{ON}/I_{OFF}	unitless
Dielectric information	Two or three of: C_{OX}, K, t_{ox}	$F(cm)^{-2}$, unitless, μm
Frequency of gate dielectric characterization	f	Hz

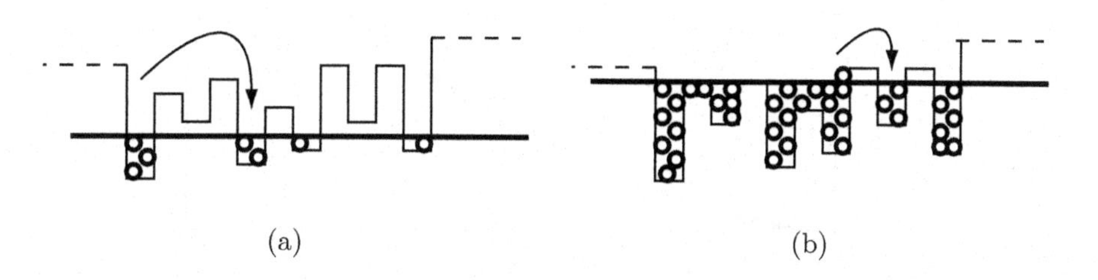

Fig. 6.17. A schematic depiction of hopping carrier transport as states are filled. In (a), where the device is less accumulated, mobile carriers populate the deeper states and see a higher barrier to movement. In (b), as carriers fill the lower states, more carriers are near the mobile carrier edge and are able to move with greater ease. The apparent mobility is a function of the accumulated carrier density.

This has two consequences. First, the threshold turn-on is more gradual than in a crystalline device because the mobile carrier density is not a Fermi distribution of carriers convolved with a step (which yields, at least in theory, a 60mV/decade subthreshold slope); it is convolved with a more gradual DOS which gives a shallower turn-on. Second, as more carriers are added, those carriers are more mobile. This means that the effective mobility is a function of the gate field and not a single modelable parameter [113]. As an example, Fig. 6.18 shows the effective mobility extracted in a set of pentacene devices using the linear region mobility extraction technique as a function of gate bias with the contact resistance compensated for [117]. Across several devices with different L values the observed mobility varies dramatically as a function of the gate bias.

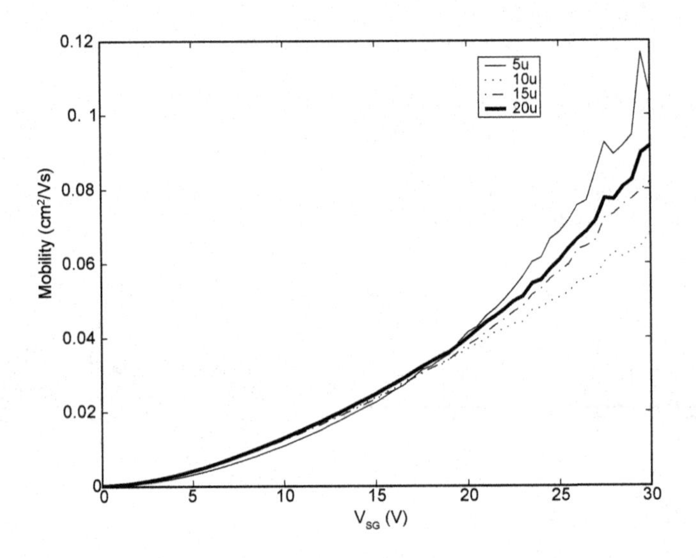

Fig. 6.18. A graph (from [117]), which shows the dependence on mobility of gate bias (and therefore gate field) on a number of transistors of different length measured using the carrier speed-mobility relationship. The effect of contact resistance has been removed in these devices to show the scaled mobility in the channel.

The net result is that curve fitting to c-Si models does not necessarily create models with strong predictive power, especially in the linear region of operation. This is a limitation of the basic curve fitting approach that many investigators are looking to eliminate through advanced characterization and modeling strategies (for example [128]).

6.5.2 Measurement and reporting

Creation of an accurate device model using these transistor parameters in many cases requires more data than is included in a single number. Mobility and contact resistance are strong functions of gate bias and need to be parameterized and included as functions of gate bias. Measurements for these parameters should be taken in the linear region, where the gate field is nearly uniform. Current experimental evidence indicates that there is no significant V_{DS} bias effect on these parameters. This is not surprising–the drain-source field is typically three or four orders of magnitude smaller than the gate field.

An ideal measurement report might include:

- Transfer and output characteristics
- A QSCV measurement (and extracted charge density as a function of voltage)
- A subthreshold characteristic
- Contact resistance as a function of gate bias (measured in the linear region)
- Mobility in the linear region as a function of gate bias

A model describing the behavior and parameters in all regions of operation can then be extracted using the techniques in this chapter.

A balance needs to be struck between using simple parameters (which may not be accurate) and use of large quantities of measured data to describe a device (which is less useful as a model). In this approach the charge, mobility, and contact resistance are measured as a function of gate voltage in a regime valid for a range of V_{DS} values. This allows the construction of a linear region models that are valid for a range of device geometries and source/drain biases. Creating a small signal model in any region of operation is straightforward after measuring g_m and r_o at the operating point of interest.

6.6 Conclusions

Characterization and modeling of OFETs is essential to understanding their behavior and effectively using the devices in circuits. The legacy of modeling in crystalline, amorphous, and polycrystalline silicon is both a helpful starting point and an obstacle to making truly accurate models. While the macroscopic behavior in all of these transistors is superficially similar, differences in the microscopic transport mechanisms complicate direct translation of models between technologies. Experimental observations and theoretical considerations have allowed a greater understanding of how OFETs work and how best to describe their behavior.

A tension exists between use of too much meaurement data, where in the extreme case polynomial fits or look-up tables might be used, and oversimplification. Models based only on measured data make circuit design and model building unwieldy and complicate intuitive circuit design, whereas oversimplified parameterized models don't describe the device well enough to be useful. This chapter presents an approach that attempts to strike a balance between the two and be both accurate and useful.

7

OFET applications

OFET devices have a number of potential applications which can take advantage of their properties. Because organic semiconductors are fully satisfied van der Waals solids which do not require any epitaxial templating, most OFET processes have a low thermal budget, simple manufacturing processes (including partially or totally printing-based process flows), and are compatible with a range of substrates. It is these applications which drive interest in OFETs, and the technology's longevity will be determined by its ability to address the challenges of applications of interest.

7.1 Displays

Direct view displays are currently the leading application for large area electronic devices. Amorphous silicon (a-Si) FETs dominate active matrix liquid crystal display (AMLCD) backplane architectures because they have a low enough process temperature to allow for economical fabrication on large glass substrates while delivering adequate performance for LCD field driving. Improvements in processing and panel architecture have allowed the commercialization of large (meter-scale) panels and the development of highly efficient manufacturing operations using motherglass sizes of extraordinary size.

Field-driven display elements such as liquid crystals and electrophoretic display materials (of which e-Ink is the leading example) are typically matrixed to create large panel sizes with a manageable number of external contacts. For a 1,000,000 pixel display, for example, a matrixed panel can use 2,000 contacts; 1000 rows and 1000 columns.

This reduction in the number of contacts comes, however, at a price. As the matrix increases in size (say, to n rows), the fraction of the frame time dedicated to each line decreases to $1/n$. The rest of the frame time ($(n-1)/n$), each row sees signals intended for other pixels. This degrades the selection ratio and reduces the field contrast available for effecting the desired electro-optical response [129]. One solution to this problem is to introduce a latching element, such as a transistor, to hold the desired field through the whole frame period and increase the fraction of the frame period the device is being deliberately driven. Other nonlinearities, such as diode responses or tunneling structures, may also be used.

OFETs are an excellent candidate for this class of backplane applications. OFETs share many of the properties of a-Si LCDs, with an even more generous thermal budget. This reduced thermal budget allows fabrication on glass, metal foils, and plastic sheet. Several groups have demonstrated mechanically flexible field-driven displays using organic field effect transistors on plastic substrates including a mechanically flexible polymer dispersed liquid crystal display [130], and displays containing e-Ink [131]. Field driving is straightforward to implement using the single transistor row/column latch architecture shown in Fig. 7.1 (a). Selecting each row individually allows programming of the field capacitor through the column lines. The row can then be deactivated for the rest of the frame time to prevent crosstalk and contrast degradation while other rows are being written.

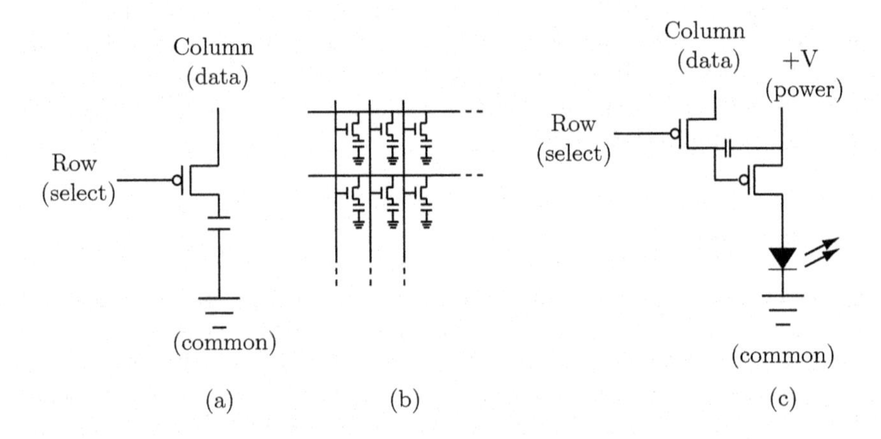

Fig. 7.1. Two architectures for display driving. (a) shows a single transistor architecture appropriate for latching and holding charge on the pixel for field driven display elements such as liquid crystal displays or e-Ink. (b) schematically shows the matrixed version of this element. (c) shows a two transistor voltage-programmed current driver appropriate for OLEDs. More advanced voltage and current programmed circuits are also possible, but require more transistors.

In addition to driving charge controlled devices such as LCDs, OFETs can also be used in current drive configurations, as shown in Fig. 7.1 (C). These architecture can be used to drive OLEDs [132], and mechanically flexible AMOLEDs have been demonstrated using OFET backplanes [133] [134]. Both voltge and current programmed circuits are possible.

In a voltage programmed drive circuit, the voltage on the gate of the drive transistor is driven directly and latched through switches on the backplane. While relatively simple to implement, this has several disadvantages:

- the scaling of voltage to current is nonlinear (approximately square law), and therefore the peripheral drive circuitry needs to have a higher bit resolution than in a linearly spaced output response
- small changes in the threshold voltage can lead to large changes in the output current

An alternative approach is the use of a current drive, in which a programming current is latched in the cell. One approach is to build a current mirror in which one arm is driven through a dummy load and then latched to the other arm (containing the OLED) each cycle. This linearizes the drive in current and also compensates for some spatial variation across the panel. Yet another approach is to drive the latching element itself and sustain the driven current. While complicated from a circuit perspective, this approach has the greatest potential for degradation compensation. Several architectures and their modeling are explored in [135].

In addition to the potential cost advantage due to easier processing via printing or evaporation, OFETs potentially offer reduced bias stress in current drive applications over a-Si transistors fabricated at less than 200°C. At these temperatures, transistors can be fabricated on a range of transparent flexible substrates and are particularly applicable to flexible OLED displays. There are also circuit and architecture advantages to using PFETS for bottom emission OLED displays [136].

7.2 Mechanical sensors

OFETs, because of their mechanical flexibility and large size, are well suited to switch and amplify mechanical actuations. The first large scale application of OFETs in a mechanical sensor was demonstrated by Someya, et al. [137]. A flexible OFET backplane was laminated together with a flexible conductor loaded elastomer whose resistance changed in response to an applied pressure. By switching through the transistor matrix and observing the current flow from a common power supply through the variable resistor, it is possible to create a force map for use as a flexible sensor skin.

In addition to producing the sensing array using OFETs, this team also fabricated much of the peripheral drive circuitry using OFET-based logic elements and introduced a creative architecture in which the unit dimensions can be customized by cutting several elements with scissors and attaching them with a pressure sensitive adhesive [138].

Another application of OFETs to measuring mechanical stimuli is the buffering of charge signals from large piezoelectric polymer sheet materials, such as PVDF. Extracting spatially localized information about the stimuli applied to the sheet can be challenging because the charge signal is dissipated across the parasitic capacitances between the stimulus and the location of the sensing circuitry. OFETs can serve as local transimpedance amplifiers which convert the charge signal into a current signal that can overcome this capacitance. This amplification can be achieved across heterogeneous substrates [139] [140] or by building the OFET directly onto the piezoelectric polymer sheet [141].

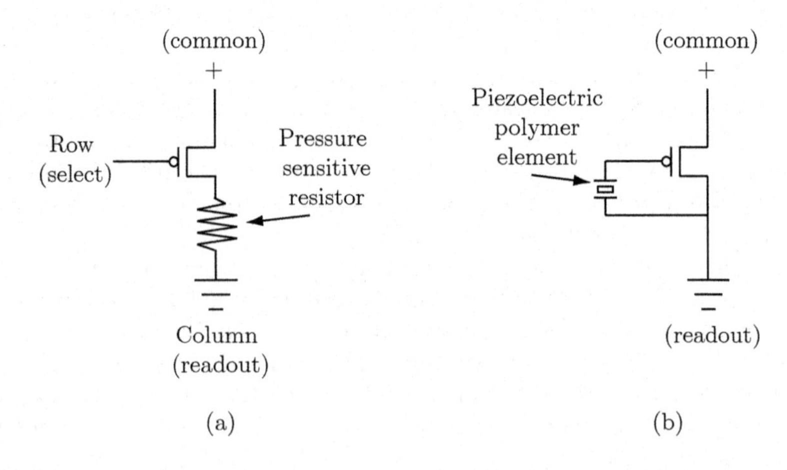

Fig. 7.2. (a) a pressure sensitive array which uses OFETs to switch an otherwise linear pressure sensitive resistor sheet. (b) shows a piezoelectric sheet film (typically PVDF or one of its derivatives) locally amplified using an OFET. The OFET converts the charge signal at its gate into a current signal which can be transmitted through a significant capacitance.

7.3 Imagers

The large and flexible backplanes available via OFET technologies can also benefit image sensing applications, many of which can benefit from large area (e.g. X-ray sensors), mechanical flexibility (contact scanners for non-planar objects) or both characteristics.

The basic architectures which have been proposed for image sensing using organic semiconductors couple a single transistor per cell OFET architecture with an organic photodiode (Fig. 7.3 (a)) [142] or a photoconductor material such as titanyl pthalocyaninie (Fig. 7.3 (b)) [143]. These architectures allow the creation of fully additive photodetector elements on essentially arbitrary substrates including plastic sheet, injection molded objects, or non-planar surfaces.

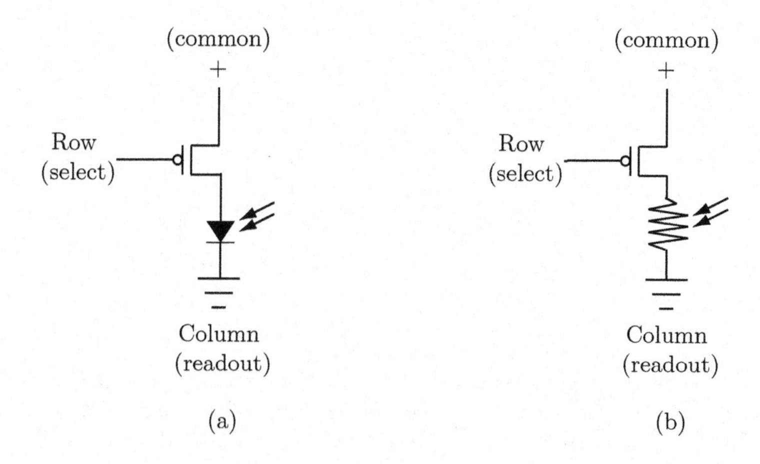

Fig. 7.3. Two imager architectures addressed using OFETs; (a) an organic photodiode as the light sensitive element, and (b) an organic photoconductor.

7.4 RFID and logic

Another potentially interesting application of OFETs is in low cost RFID tags and logic elements. If transistors, passive elements, and the tag antenna can be produced simultaneously using continuous printing processes, it is conceivable that economies similar to those achieved in paper printing can be achieved.

More circuit details are offered in Appendix D, but structurally, the simplest RFID tags consist of an adjustable load rectifier, a code generator or counter, and logic gates to convert the count into the RFID code. Several examples have been demonstrated in the literature using printed [61], shadow masked [74], and lithographically processed [144] OFET devices. The simplest architectures use the ring oscillator output directly to generate the code, more sophisticated logic incorporates full counters and decoding circuitry.

7.5 Conclusions

The applications OFETs are applied to will continue to define the requirements for material, circuit, and device performance. The demands on OFET

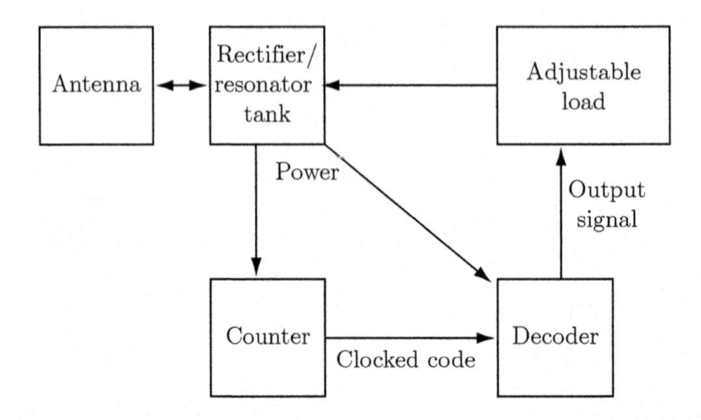

Fig. 7.4. A block diagram showing the elements of a typical organic RFID circuit. The clock generates a repeating sequence of codes, typically using a ring oscillator as a timebase. The decoding circuit analyzes the clock signal and turns the load on and off at the appropriate times, generating the output sequence.

performance in a field-driven display, for example, are quite different from those in a current driven display. It is ultimately the appeal of compelling applications which will drive continuing research, product development, and commercialization of OFET technology. It is therefore essential to understand the demands of these applications and the value that OFET technologies can bring to them.

A

Appendix A: Review articles

The following list contains several review and method papers (and is itself strongly influenced by [17]). This list is certain to be both incomplete and out of date, but may serve as a starting point for further research.

- N. Greenham and R. H. Friend, in *Solid State Physics: Advances in Research and Applications*, Vol. 49, H. Ehrenreich and F. Spaepen, Eds., Academic Press, San Diego, 1995, pp. 1149.
- A. J. Lovinger and L. J. Rothberg, "Electrically Active Organic and Polymeric Materials for Thin-Film-Transistor-Applications," *Journal of Material Research* 11, 1581, 1996.
- H. E. Katz, "Organic Molecular Solids as Thin Film Transistor Semiconductors," *Journal of Material Chemistry,* 7, 369, 1997.
- A. R. Brown, C. P. Jarrett, D. M. de Leeuw, and M. Matters, "Field-Effect Transistors Made From Solution-Processed Organic Semiconductors," *Synthetic Metals* 88, 37, 1997.
- F. Garnier, "Thin Film Transistors Based On Organic Conjugated Semiconductors," *Chemical Physics* 227, 253, 1998.
- G. Horowitz, "Organic Field-Effect Transistors," *Advanced Materials* 10, 365, 1998.
- H. E. Katz and Z. Bao, "The Physical Chemistry of Organic Field-Effect Transistors," *Journal of Physical Chemistry B.*, 104, 671, 2000.
- C.D. Dimitrakopoulos and P. R. Malefant, "Organic Thin Film Transistors for Large Area Electronics." *Advanced Materials*, 14, 99, 2002.
- C. D. Dimitrakopoulos and D. J. Mascaro, "Organic thin-film transistors: A review of recent advances," *IBM Journal of Research and Development*, 45, 11, 2001.
- G. Horowitz, "Organic thin film transistors: From theory to real devices," *Journal of Material Science*, 19, 1946, 2004.
- G. Malliaras and R. Friend, "An organic electronics primer," *Physics Today*, 58, 53, 2005.

- C. R. Newman, C. D. Frisbie, D. A. da Silva Filho, J. L. Bredas, P. C. Ewbank, and K. R. Mann, "Introduction to organic thin film transistors and design of n-channel organic semiconductors," *Chemistry of Materials*, 16, 4436, 2004.

B

Appendix B: Recipes and equipment

This chapter contains recipes for a reference OFET process flow. It is intended as a starting point for experimentation and further development. The names of specific products and equipment which have been tested are included, but are not intended to exclude alternatives from consideration.

Two important resources in the literature for general semiconductor etch recipes are [62] and [63]. These two papers provide a wealth of wet and plasma etch recipes and rates, including the differential attack rate of some etches against unintended target materials.

B.1 Overall process flow

The overall process flow which will be laid out is based on [76], with some further refinements.

B.2 Notes on equipment

Any process needs to be adapted to the equipment available. Ideally, OFET processing can be performed in a cleanroom which is equipped with good facilities for cleaning, etching, photolithography, and deposition.

Cleaning and drying are especially critical operations. High purity deionized water should be used for the final rinse on all steps, and ideally substrates will be dried in a spin/rinse dryer. If a spin-rinse drying unit is not available, a chemically resistant spin-coater can be configured for spin etching, rinsing, and drying as an alternative (or just rinsing and drying). Drying with a nitrogen gun, while possible, is not ideal. The nitrogen stream from guns often contains particles and drying is usually not perfect, which runs the risk of material drying onto the substrate.

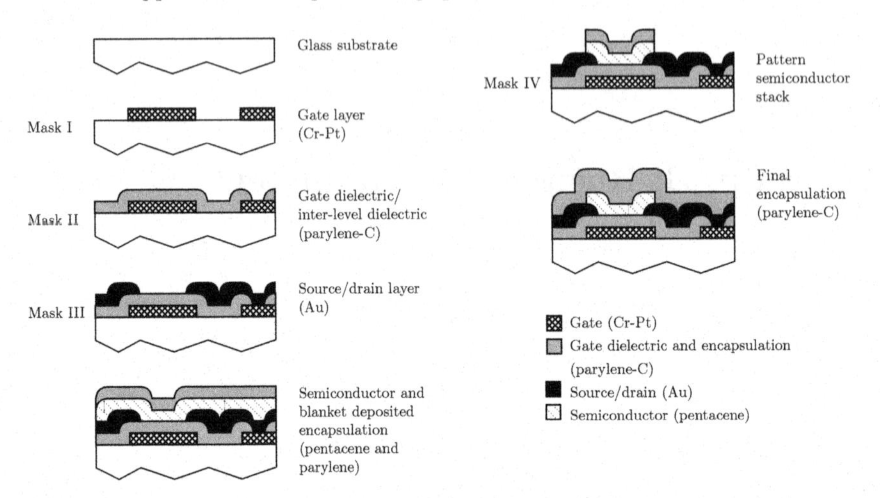

Fig. B.1. The process flow in Kymissis, et al. [76]. All layers are subtractively etched, and parylene is used to protect the pentacene from attack by the photoresist and developer.

B.3 Substrate preparation

The first step is to select a suitable substrate for preparation. The substrate will depend on the application, for development work glass is a typical choice. Inexpensive soda lime glass is available cut and cleaned from a number of sources, and is usually adequate after some additional cleaning but has a non-negligible surface roughness. Sheet drawn display glass, polished float glass, and polished sliced glass wafers are more ideal choices.

The cleanliness of the substrate is essential; particles are often incompletely coated by dielectric layers and can cause shorting. Surface oils and soil will also cause defects in the conductor and insulator layers.

The cleaning process should be tuned to address the surface contamination present. For typical pieces which have minimal glass grit, a straight clean of 30 minutes in piranha cleaning solution (1:3 concentrated H_2O_2:H_2SO_4) is typically adequate for removing organic residue and light particulate contamination.

Glass particles are difficult to remove. Etching of fine particles with KOH, dilute ammonium fluoride, or HF is possible, but difficult and hazardous, and can roughen the substrate. It may be best to switch glass suppliers or request better cleaning of the cut glass if there is significant glass particle contamination.

High grade plastic foils which have no lubricating films or particles and are 125 microns (5 mils) and thicker can easily be used freestanding and temporarily mounted on glass with a few drops of water during photolithographic operations. Temporary mounting of such foils onto glass or Si carriers is also possible; one option is to use a spin coated layer of PDMS, another is to spin coat a debondable adhesive (e.g. Dymax OP-18). Plastic sheet can be cleaned using the following protocol:

- 30 minutes ultrasonic agitation in detergent (e.g. 1% Micro90 in deionized water)
- Rinse
- 5 minutes ultrasonic agitation in deionized water
- Rinse
- 5 minutes ultrasonic agitation in deionized water
- Blow or spin dry

B.4 Gate deposition

The gate is the layer which is best bonded to the substrate, and therefore needs to be well attached to survive external probing and/or other attachment. If wire bonding or heat seal connector bonding is going to be used, the surface needs to have a composition and surface condition suitable for those operations.

A typical gate is 5nM of chromium followed by 100nm of gold thermally or e-beam deposited. Other materials can be used, but on glass especially, an adhesion layer of a reactive metal (e.g. chromium or titanium) should be used for scratch resistance. If the substrate will be probed, using 100nm of chromium or aluminum is suitable. Thick gold should be used for gold ball bonding, and aluminum should be used for aluminum wedge bonding. If a self-aligned process is to be used, the 5Cr/100Au gate is too transparent–at least 50nm of chromium should be used to increase the optical density.

Use of HMDS on gold is unnecessary. HMDS use is advisable on any metal that forms an oxide (e.g. aluminum or chromium).

After photolithographic pattern definition (which is covered in more detail in the next section), gold can be etched in a 30% solution of KI in DI water, then rinsed. The chromium adhesion layer can be etched using any of several etching cocktails [62] [63] or pre-mixed commercial reagents (e.g. Transene Chromium Mask Etchant), then rinsed and stripped.

Photoresist stripping can be performed using any of several commercially available stripping solutions (e.g. Microstrip 2001 from FujiFilm electronic materials). Samples should be immersed in the stripper for 10 minutes (using agitation if necessary), immersion in a fresh bath for another 10 minutes, then the samples should be rinsed with DI water and dried. While acetone will remove most photoresist, it is not an ideal choice for photoresist stripping because of its high volatility which runs the risk of residue redeposition. NMP-based stripping solutions are a better choice.

Because the initial surface is glass, the gate can be deposited using sputtering and plasma processes can be used to remove the photoresist instead of solution processes,

B.5 Photolithography

Photolithography can be performed using chromium or emulsion masks. Chromium masks can hold tighter resolution and alignment tolerance and remain cleaner than emulsion masks, but quality emulsion masks are available which can be used down to 4 microns or better resolution.

A typical photolithographic process is as follows (for positive resists such as FijiFilm OCG 825, unconverted Clariant AZ5214, or Shipley Microposit 1813) :

- Prime substrate if necessary (ideally in wafer prime oven, otherwise spin coat on a liquid primer)
- Dispense resist at 500RPM
- Ramp to 750 rpm for 6 seconds to spread the resist
- Spin for 30 seconds to thin the resist to 1-2 microns 3000 rpm for OCG825, 5000 rpm for 1813 and 5214
- Bake resist for 30 minutes in a 90°C oven or for 120 seconds on a 90°C hotplate
- Expose resist through mask
- Develop in developer
- Rinse and dry pattern
- Post-bake if necessary (for metal and parylene etching it is usually not necessary or advisable to post-bake). To post-bake place substrates into a 110C oven (30 minutes) or hotplate (2-5 minutes).

If thicker photoresist is desired reduce the spin speed (not the spin time) or change to a higher viscosity formulation.

B.6 Gate dielectric

Parylene-C is a convenient and straightforward gate dielectric to use. Parylene-C dimer (and other variants) is available from a number of suppliers including Specialty Coating Systems and Uniglobe Kisco (dix-C). Approximately 200nm of gate dielectric material provides a good starting point for many circuits with operating voltages in the 5-20V range. Many other options are viable depending on the equipment available and application requirements. Photoimageable polyimide (e.g. HD Microsystems 8820) is a straightforward alternative if a parylene coater is not available.

Typical parylene deposition conditions in a Specialty Coating Systems LabCoater 2010 are:

- 0.4g initial charge
- Pyrolysis at 690°C
- Deposition pressure at ‘50’ (which is about 50mtorr)
- Maximum vaporizer temperature at 180°C

Once deposited parylene-C can be patterned using regular photolithographic processes. Priming is not necessary, photoresist spins well on parylene-C. Etching is accomplished in an oxygen atmosphere RIE. Parylene gate dielectrics can also be etched in a barrel asher. While 10-20% overetch is usually desirable (if the chamber etch rate is not uniform even more overetch may be warranted), the photoresist will begin to polymerize and become more difficult to remove. Should this occur, thicker photoresist can solve the problem; an unpolymerized layer remains in contact with the substrate which can be stripped.

Parylene is not covalently bonded to the substrate in a normal process, and is held on by van der Waals forces. If greater scratch/delamination resistance is required, the surface should be primed with γ-methacryloxypropyl trimethoxysilane or vinyl trichlorosilane to anchor the parylene. Parylene covalently bonds with the end groups of these silanes and the resulting parylene is significantly better attached to the substrate. Use of air guns for blowoff is not recommended; the focused air jet can sheet off parylene layers. Spin processing using a spin coater or spin rinse dryer works without trouble.

B.7 Source/drain

Gold sticks reasonably well to parylene-C. It can be evaporated onto the gate layer and lithographically patterned without the need for a separate adhesion layer. Thermal or e-beam evaporation is strongly recommended; the active ions created during sputtering can damage the gate dielectric surface. 60nm of gold is typically adequate, and is transparent enough to use in a self-aligned process as well as a traditional process. Photoresist residue present after gate layer patterning is highlighted by the deposited gold and can interfere with adhesion.

B.8 Semiconductor layer

After stripping and cleaning of the source/drain layer, the semiconductor can be deposited. Purified pentacene can be evaporated on the prepared substrate directly. The exact optimal deposition conditions depend on a number of factors including the system configuration and the systematic errors in temperature and rate metrology due to geometrical factors, etc. As a starting point the substrate should be held at 60°C and the rate should be 0.01nm/s.

Regioregular poly(3-hexyl thiophene) should ideally be handled under an inert atmosphere (for example in a nitrogen or argon filled glovebox or, less ideally, in a glove bag system). Dissolve 0.1% by weight OFET-grade regioregular poly(3-hexyl thiophene) in anhydrous chloroform, spin overnight to dissolve. Filter twice through two new 0.45μm filters, then puddle on substrate and spin at 1500rpm. The solution will dry on the substrate. The polythiophene can then be coated with parylene and lithographically processed. Ideally, photoresist spin and bake are performed in the glovebox. Drop-cast and printed polythiophene may need to be prepared at a higher concentration to achieve reasonable results; the concentration needed requires some experimentation. Chloroform wets parylene-C especially well and another gate dielectric (e.g. cross-linked polyimide) may be a better choice for defined area printing such as inkjet printing if small features are required.

Both layers, if blanket coated, can then be coated with another 200nm of parylene-C and patterned directly. Stripping the photoresist is not recommended. An additional sealing with parylene can be used for further protection as well. A vapor barrier (for example thin gold or sputtered SiO_2) can then be deposited on the parylene for further stabilization.

B.9 SAM treatment

Once a baseline process is established, processes for treating the electrodes and gate dielectric can be explored. To treat a surface with most SAMs (e.g. nitrobenzenethiol, octadecyltrichlorosilane, etc.), a non-reactive beaker should be used (i.e. use plastic beakers with silanes, glass or plastic is suitable for thiols). A 1-0.1% solution of the SAM should be mixed in a suitable alcohol (typically ethanol) and the sample immersed for at least 10 minutes in the solution (but ideally overnight). For overnight runs the container should be covered with a watch glass or plastic film (not paraffin-based film) to reduce evaporation. Gold has a tendency to delaminate during thiol treatment; it is often necessary to anchor it using an adhesion layer.

All of these materials should be handled in a fume hood with proper personal protective equipment. Note that thiols are especially malodorous and, even with proper precautions, their scent can adsorb onto the surface of gloves and clothing which enters and exits the fumehood environment and enter the room. Leaving used gloves, pipettes, washed glassware, and arm shields in a hood overnight before disposal can reduce scent carry-over.

B.10 Process summary

The attached chart summarizes the steps used in the process flow described in this chapter (using pentacene deposition as the example). High purity water should be used for all rinse steps.

Table B.1. Example pentacene process

Layer	Operation	Parameters
	Piranha clean	30 minutes in fresh, hot solution
Gate	Gate evaporation	5nm Cr/100nm Au (e-beam or thermal)
	Photo	Gate mask
	Etch	KI, rinse, chromium etchant, rinse
	Strip	Soak in stripper 10 min, rinse in water
		soak in fresh stripper 10 min, rinse in water, dry
Dielectric	Deposit parylene-C	200nm layer
	Photo	Via mask
	Etch	oxygen plasma
	Strip	Soak in stripper 10 min,
		Soak in fresh stripper 10 min, rinse, dry
Source/drain	Source/drain evaporation	60nm Au (e-beam or thermal)
	Photo	S/D mask
	Etch	KI, rinse
	Strip	Soak in stripper 10 min, rinse
		Soak in fresh stripper 10 min, rinse, dry
Active	Deposit pentacene	60°C substrate, 25nm total at 0.01nm/s
	Deposit parylene-C	200nm layer
	Photo	ACT mask
	Etch	oxygen plasma
Encapsulation	Deposit parylene-C	200nm layer

C

Appendix C: Device layouts

This appendix offers several suggestions for device and mask layout. It is written for lithographic processes, but many of the concepts also apply to printing and other patterning techniques.

C.1 Design rules

C.1.1 Pattern and overlay parameters

One of the first considerations when laying transistors out is the applicable design rules. These will be determined by the yields required and the performance of the process in question.

Three parameters are of particular significance:

1. What is the precision with which overlapping layers can be aligned (α)?
2. What is the finest gap that can be produced (β)?
3. What is the finest line width which can be produced (γ)?

For simplicity these will be referred to as α, β, and γ. These should be selected so that they can produce the required yields.

C.1.2 Test mask for pattern and overlay parameters

The simplest way to evaluate a process step is to create a test mask which directly measures the process parameters.

An array of patterns like the one in Fig. C.1 at different sizes can be used to test photlithographic resolution. Testing on both the horizontal and vertical axes (and on diagonals if diagonal runs will be used) can identify directional process bias (e.g. shear forces from etch tank dipping or spin/rinse drying) which need to be addressed. The checkerboard pattern helps both to see when the etching is complete and to determine the degree of under/over exposure

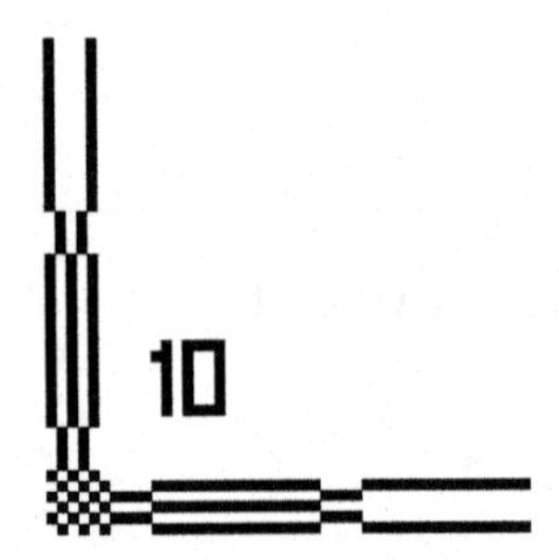

Fig. C.1. A sample resolution test pattern.

and etch. As with all test structures, interpretation is greatly simplified if the test structure size is printed along with the test structure itself.

The precision of overlap can be measured using verniers, shown in Fig. C.2. These need to be combined with alignment marks (see Section C.2). Verniers allow measurement of offsets which are smaller than the smallest structure which can be printed. Two ruled elements are created (one on each layer of interest) with a slight difference in their pitch. If the two are perfectly aligned (i.e. to better than the difference in spacing Δ) then the middle rules will align. If the misalignment is equal to Δ, the first rule off of the center will align, and so on. At least one horizontal and one vertical vernier for each layer pair is recommended. By measuring the relative offset of multiple vernier sets with a known distance angular misalignment and substrate warpage can also be measured.

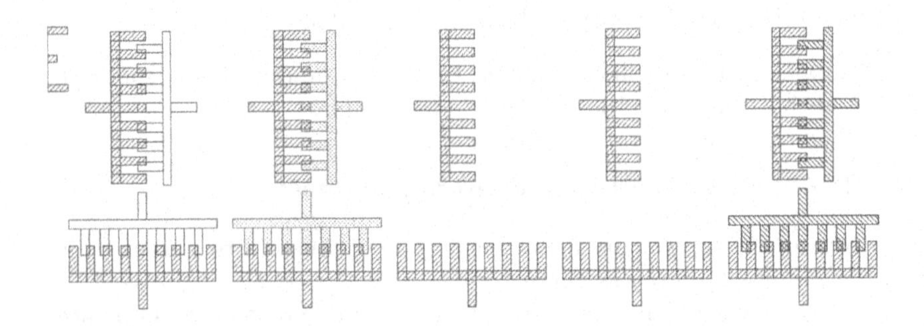

Fig. C.2. Sample verniers for measuring photolithographic misalignment. The first pattern is printed with the gate and the matching patterns are printed on subsequent layers. Both vertical and horizontal verniers should be used.

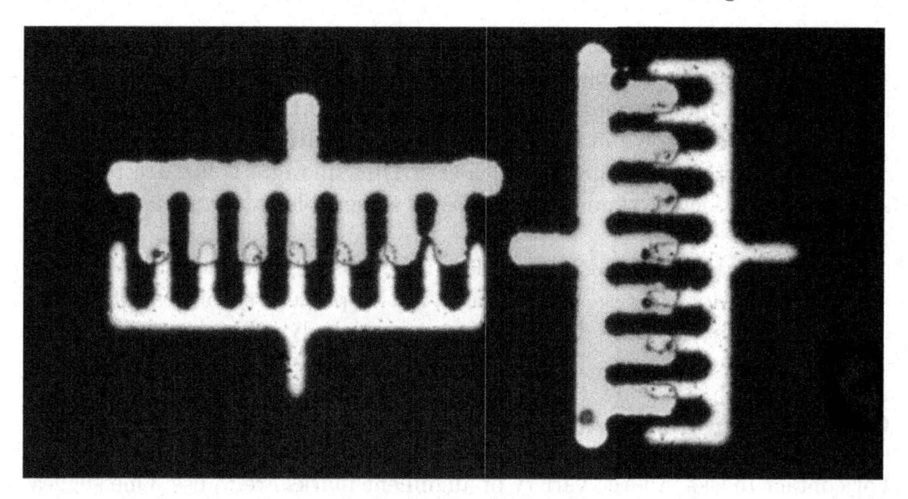

Fig. C.3. A photograph of verniers indicating the vertical and horizontal alignment between layers.

The precision with which alignment can be achieved, α, is a function of the equipment used, the alignment marks used, and the skill of the operator. Monitoring of the misalignment can help isolate process problems when they occur, and also allows appropriately aggressive design of the mask so as to improve device performance.

C.1.3 Determining design rules

A series of experiments should be performed with the test mask set to determine the minimum values for each of these three parameters achievable. A series of samples should be prepared, ideally with optimization for exposure, dose, and development time for photolithography first and etch optimization second. This will maximize the fidelity of the photolithographic pattern.

Standard process recipes are useful as a general guideline for photolithography, but the reflectivity of the underlying layers can play a significant role in the optimum dose. Gold, for example, will expose significantly faster than indium tin oxide on glass. This should be optimized to avoid over or under exposure. Etch solution concentration and etch time should be determined for wet etchants at the pattern density of interest; etch power, chemistry, and time for plasma processes. Of particular concern for plasma processes is optimization to reduce the formation of polymerized resist on the surface, which can be challenging to remove.

Once a somewhat optimized flow is established, α, β, and γ should be measured across a representative sample set. Based on the mean, observed variation, and acceptable yield, the design rule should be selected. For a typical process, approximately 3 standard deviations more conservative than the average should be selected. Selecting design rules (especially α) that are too conservative can waste space and sap performance (since there will be too much overlap), but without device and circuit yield there is no performance.

The examples shown in this section are conservative and are based on use of a SVG EV-1 contact aligner. Although 2μm line and space are routinely achievable and better than 1μm overlay is achievable, α=5μm β=10μm and γ=10μm.

Even after a device mask set is designed, these marks should be retained and measured for process monitoring. As the process stabilizes and yields improve, the mask can be designed more aggressively for high performance.

C.2 Alignment marks

For contact masks, a large variety of alignment marks are in use. One suggestion is shown in Fig. C.4. The marks should be large enough for the equipment of interest, and the gap between the box in square and cross in squares should be on the order of α. The four layers shown in Fig. C.4 are the gate (unlabeled), the via layer (VIA), the source/drain layer (S/D), and the semiconductor layer (ACT).

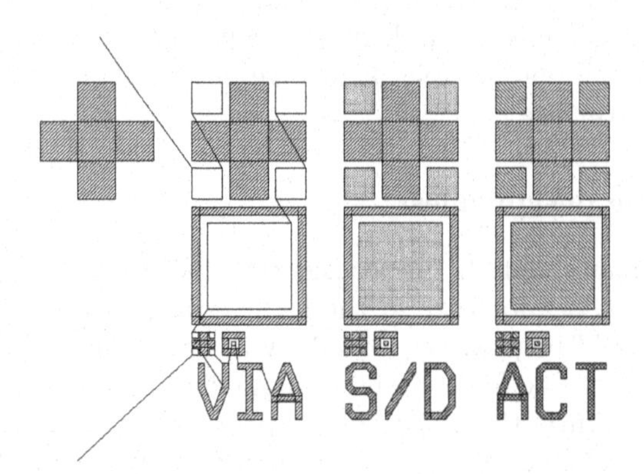

Fig. C.4. An example set of alignment marks including both high and low resolution cross/square and box-in-square types. The base pattern in printed with the gate layer and subsequent layers are aligned to the base pattern. The via layer is shown only in outline. Because the via layer is dark field, the via alignment marks need to be inverted so that the patterns are visible during alignment when a positive resist mask is used. One approach to doing this is to draw the alignment marks in the dark field layer, and then XOR the whole via alignment mark area with a large rectangle before or after inverting the entire via layer pattern.

It is generally preferable to print all of the alignment marks on the first layer and then align all of the other layers to this first layer. This prevents

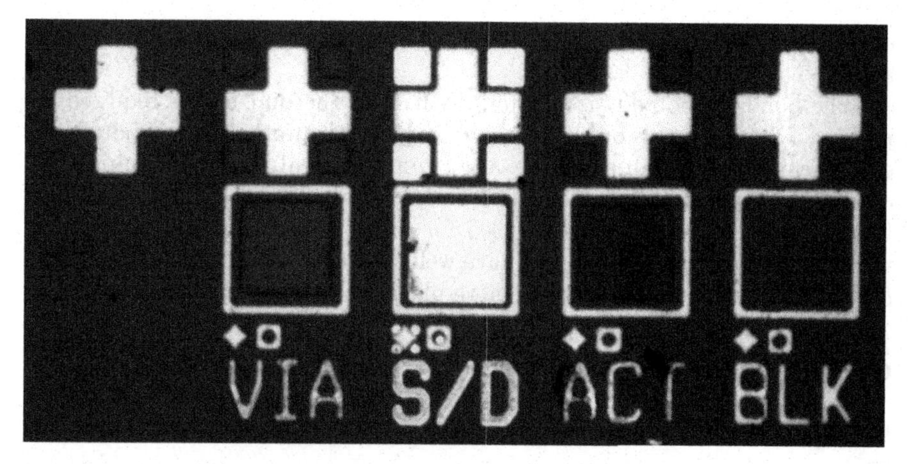

Fig. C.5. A micrograph of the same set of alignment marks shown in Fig. C.4. Note the inversion of the via layer in the final pattern.

additive errors from accumulating in the device stack. In some cases the first layer may be difficult to see (e.g. if ITO or another transparent or semi-transparent layer is being patterned). In these cases alignment marks can be put down as separate first step, on the second layer, or in regions where the alignment marks are needed can have a visible metal placed on them.

C.3 Transistor layout

Test transistors allow for process monitoring and straightforward extraction of device performance. Several material parameters (such as the dielectric capacitance and the gate source/drain misalignment) can also be measured directly from a transistor. Transistors with a range of channel lengths should be included to allow for contact parameter extraction. A detailed layout is shown in Fig. C.6.

C.3.1 Probe/bond pads

An important consideration for testing is the quality of the bond between the metal layers and the substrate. The source/drain layer especially is susceptible to scratching (if probes are used) or delamination (if wirebonding, anisotropic adhesives, or heatseal connectors are used). This is an issue for circuits at nodes which will be externally bonded or probed.

At least one of the metal layers needs to be scratch/peel resistant, and this is usually the gate layer against the substrate, although other combinations are possible (e.g. source/drain layer can also bond well to the gate dielectric and be deposited over a via). A typical solution is to place a via in the center of the bond pad and leave both gate metal and source/drain metal over those areas.

The bond/probe pad size should be at least the dimensions required for probing, which depends on the system being used to probe. Manual bonding is easily achieved in a $100\mu m$ square, and $80\mu m$ is typical. Manual manipulators can probe significantly smaller structures, typically $10\mu m$ square is straightforward and larger pads can be used for convenience. Heat seal connectors vary significantly in size, but will have well specified design rules. A border of α should be left around the pad before placing any other critical devices.

C.3.2 Channel layout

For test transistors the following guidelines can be followed:

- The channel length L should be at least β.
- The border of the active area should be at least α away from the source/drian electrodes. While the device will function if the active area is smaller than this (it will be fine even if it is shorter than the source/drain electrodes), it is advantageous to have it extend α beyond the electrode edge. This provides the minimum source/drain overlap capacitance and the largest effective W for a given actual W.
- The gate should extend beyond the source/drain electrodes 2α. This avoids having uncontrolled semiconductor channels connecting the source and drain.
- The source and drain should overlap the gate at least α. Too little overlap will lead to some devices having ungated channel segments, which introduces a large series resistance and is problematic. Too much overlap increases the parasitic capacitance without improving performance.
- All vias should be at least α away from the closest semiconductor edge. This prevents gate leakage through the semiconductor material into the via
- The electrode width must be at least γ
- The transistor width must be at least γ
- The gate line in the transistor length dimension must be at least γ
- All probed sites should have the correct bond pad design and be spaced to permit probing using the intended system

Ideally, all of the transistors (even those of different length and width) will have the bond pads in the same location. This simplifies testing of arrays on most probe stations by lifting the platen, moving the sample, and dropping it without moving the probes. A probe card can also be used to increase efficiency.

An example layout is shown in Fig. C.6, an example array in Fig. C.7, and a micrograph of a test transistor is shown in C.8.

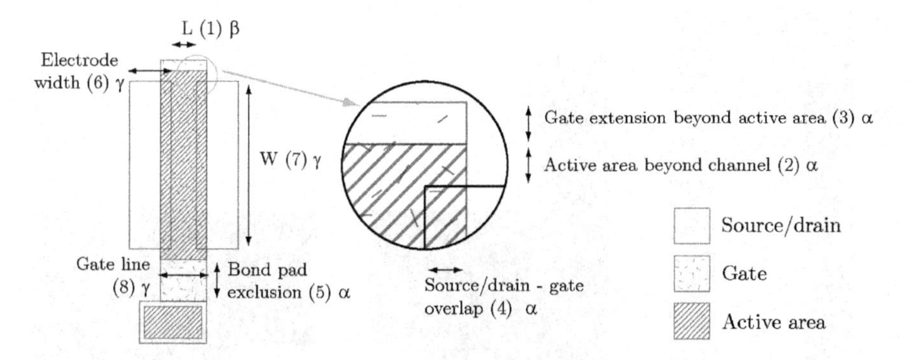

Fig. C.6. The suggested design rules for an OFET. While the active area can be set at virtually any size, there are some minor advantages to extending it beyond the source/drain electrode.

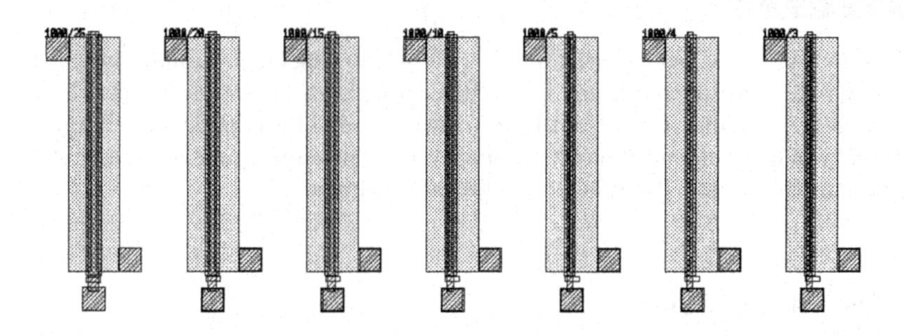

Fig. C.7. An example transistor array. Testing throughput can be improved by placing all of the bond pads in the same relative locations, which allows testing of multiple devices without the need to rearrange the probes or probe card.

C.3.3 Circuit layout

Devices used in circuits have one major difference with test devices– there are usually many fewer bond pads in the design, and therefore less opportunity for gate leakage into the device because of vias to the gate layer. The same guidelines apply, and a forbidden zone at least α wide should be respected around vias.

C.4 Transfer line

The transfer line is a series of transistors with different channel lengths. It is not necessary to create a separate test structure for this purpose if individual transistors of varying channel length are provided, but Fig. C.9 (a) shows an example dedicated transfer line.

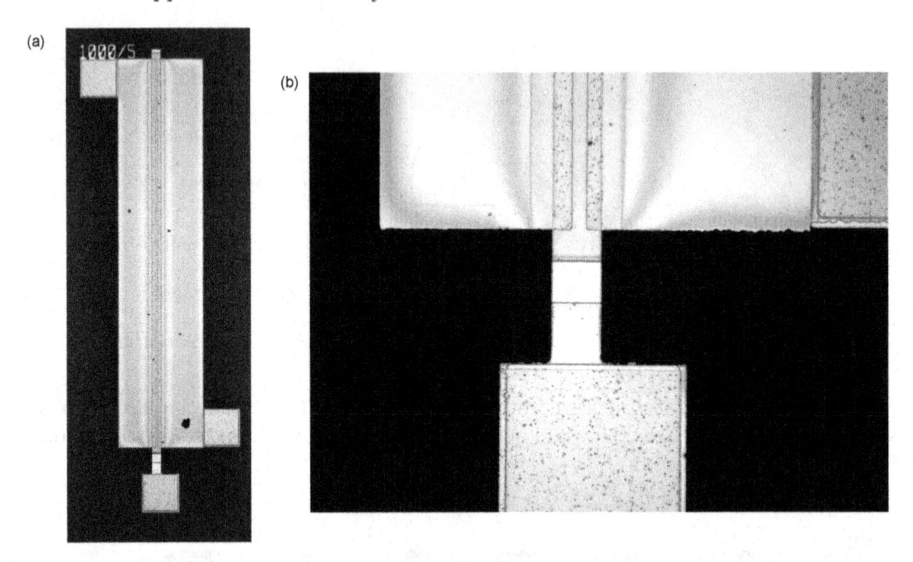

Fig. C.8. (a) A typical test transistor, (b) a detail showing the overlap between the active area, gate, and source/drain. The large contact pads at the source, drain, and gate have large vias and metal from both metal layers. This significantly improves the reliability of contacts made using probing systems when the gate is more scratch-resistant than the source/drain layer on the gate dielectric.

C.5 Test structures

In addition to transistor arrays, several test structures are recommended for device wafers. These are summarized in Fig. C.9.

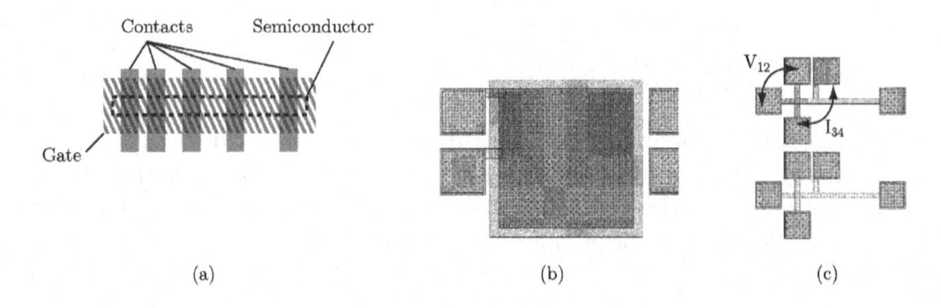

Fig. C.9. Several test structures for determining material parameters. (a) schematically shows a transfer line, which can be useful for extracting contact resistance. (b) shows a resistance test structure, and (c) shows a capacitor which can be used to determine the capacitance of the gate dielectric.

C.5.1 Resistivity

A van der Pauw structure can be used to determine the sheet resistance of conductor layers. Since the structure is symmetrical, the sheet resistance R_s is given by $R_S = \frac{\pi R_{12,34}}{ln2}$ where $R_{12,34} = \frac{V_{12}}{I_{34}}$ as indicated in Fig. C.9 (c). An example structure is shown in C.10.

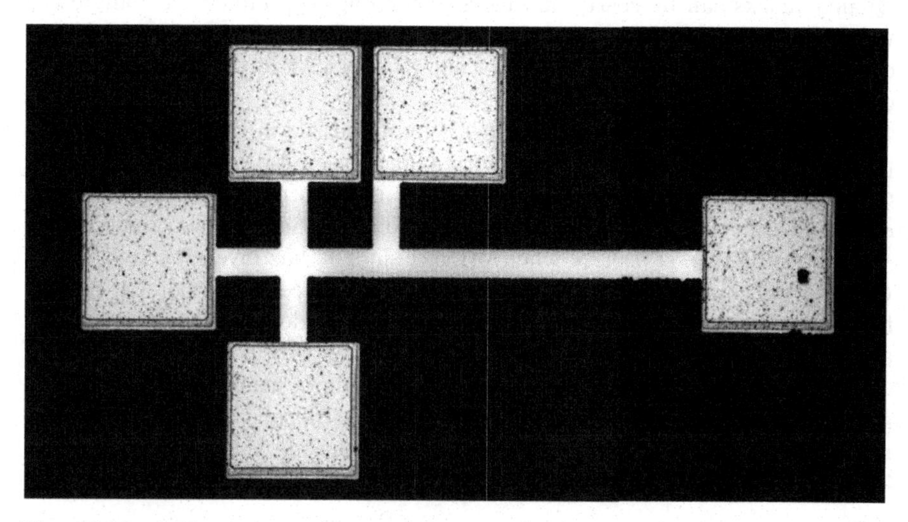

Fig. C.10. A sheet resistivity and linewidth test structure. The sheet resistance can be measured in the van der Pauw feature, and the resistance of the line can be measured using a two, three, or four point technique. The bar is nominally 10 squares, so the degree of overetch can be quantified.

C.5.2 Capacitance

While it is possible to measure the gate dielectric capacitance through QSCV overlap, it is advisable to include several metal/insulator/metal capacitors to allow independent measurement of the gate dielectric. The capacitors can be formed using the gate and the source/drain layers.

C.5.3 Optical thickness

If an ellipsometer or interferometer is to be used to measure layer thickness, large pads on the die or at the substrate level need to be provided. Monitor samples can alternatively be used to conserve device area. Transparent materials and thin metals on large monitor samples can also be examined with a UV-Vis system (the $\lambda/4$ peak will appear for transparent materials with an index constrast to the substrate, metal thickness can be gauged by optical density). Scan lines for profilometer, SEM cross-section, or AFM analysis are also advisable if use of those techniques is anticipated.

C.6 Quad mask design

Quality masks can be costly and time consuming to produce. In many cases where not all of the substrate area is required for testing, it is possible to combine four masks onto one plate to allow faster prototyping and economize on mask production.

One strategy for mask consolidation is summarized in Fig. C.11. The layer pattern is separated into four dice, with a 90 degree rotation around the same common vertex for each successive layer. This pattern is then tiled over the substrate area. The mask is then used rotated by 90° for each successive layer. $\frac{3}{4}$ of the substrate is useless when printed since it has layers printed in the wrong order, but $\frac{1}{4}$ of the area comes out with the correct order. Of course, four more layers or design options (e.g. self-aligned devices) can be added to a second mask.

Lateral shift can also be used to consolidate more mask layers. In contact aligners, this approach has some disadvantages. Because the pattern lacks rotational symmetry, the die size is limited to the mask movement distance in the alignment system used.

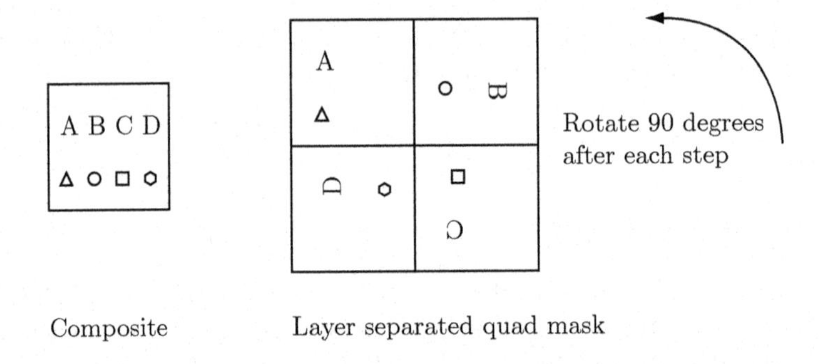

Fig. C.11. One strategy for consolidating multiple layers onto a single mask. Using rotation instead of lateral translation allows a wider range of die sizes to be used.

D

Appendix D: Logic gates

Many of the applications using OFETs use logic elements to implement logic functions or store state. Row decoding, display state maintainance, and code generation for RFID all use logic elements to implement their functionality. Logic gates made using OFETs are typically made using PMOS, NMOS, or pseudo-PMOS/pseudo-NMOS logic architectures. These design approaches typically use only one type of transistor and are well suited for processes which use a single semiconductor.

D.1 Inverters

The simplest logic gate, the inverter, is the basis for the design of other more complicated logic gates. A high voltage applied to the input is output as a low voltage, and a low voltage as a high voltage. If a logical '1' is represented as a high voltage then the output is the inverted (or NOT) logical function of the input.

Three approaches for making an inverter using a PFET are shown in Fig. D.1. Fig. D.1 (a) shows the classical design for an inverter using a transistor. When the transistor is on (i.e. when the input is low), the . Of course the resistor value should be selected so as to exceed the transistor resistance when the device is in depletion, but be less than the device in accumulation. When the transistor is accumulated the transistor is the smaller resistance in the voltage divider and pulls the output value high. When the transistor is off the transistor pulls the output low. This approach can be graphically understood using a loadline analysis, shown in figure D.2.

One complication with a resistive load device is that the transistor channel resistances can be very high–typically in the megaohm or even gigaohm range–and the metals used for the gate and source/drain layers typically have sheet resistances in the 10-100 millioms/square range. Load resistors matching the transistor would occupy a large area and be impractical to make in many circumstances.

An alternative approach is to use a second transistor, biased partially or wholly into accumulation. Depending on the transistor threshold voltages, it may be advantageous to set the gate voltage to the ground reference, the positive voltage rail, the output value (which will be between the power rails, but not a constant), or to an arbitrary value (possibly even outside the power rail voltages) via an independent voltage rail. Fig. D.1 (b) and (c) show this approach, which is significantly more space efficient. If the threshold voltage does not allow an appropriate load resistance to be established, it is also possible to create the load by backgating the load transistor device to shift the effective V_T [145].

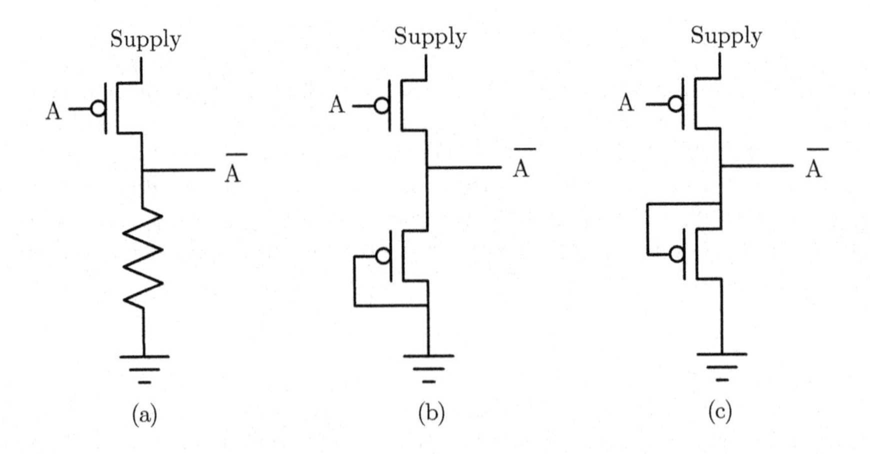

Fig. D.1. Three strategies for fabricating inverters; (a) resistive loading, (b) saturated loading, and (c) depletion mode loading. The optimal load transistor bias is determined by the threshold voltage of that transistor; biasing can also be st at the high power supply rail or to other arbitrary voltages as appropriate.

D.2 NAND and NOR gates

More complicated logic elements can be formed by placing two transistors in series or parallel to form NAND or NOR gates. Series transistors only allow the output to be pulled up when all of the transistors are on together. Parallel transistors allow the device to be pulled up when any of the inputs is on. Two architectures are shown in Fig. D.3.

One complication in the NOR circuit is that the source voltage on the transistor near the output is decreased by the voltage drop across the transistor attached to the power rain–this drop will be approximately V_T, and

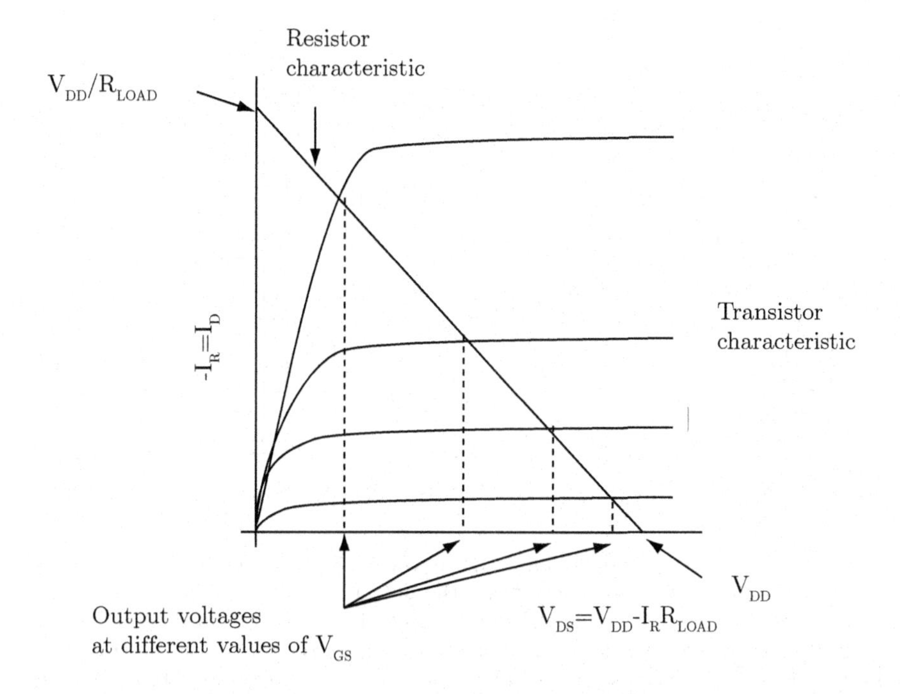

Fig. D.2. A schematic showing the load line method applied to a resistively loaded inverter architecture. This method provides a graphical explanation of simple series circuit operation. The total voltage applied across both the resistor and the transistor is a constant, and equals the positive power rail. The current flowing through both devices also matches, so by superimposing both characteristics and looking at the intersection point, the operating point can be determined. The transistor characteristic changes with changing gate voltage which leads to different operating points as a function of the gate bias. The same procedure can be applied with other loads (e.g. ca saturated load transistor or complementary transistor).

another drop of V_T is seen across the second transistor. This severely limits the output swing of the NOR gate and usually makes it impossible to design a device which exhibits voltage gain. It is for this reason that NAND gates are typically preferred in OFET circuits.

D.3 Ring oscillators

A ring oscillator is a metastable circuit formed by connecting an odd number of inverters in series and feeding back the last inverter's output to the first. Any ring oscillator which has enough elements and is made with logic-level sustaining elements will spontaneously oscillate; the loop gain is, in principle, infinite and any noise in the system will be amplified into a sustaining oscil-

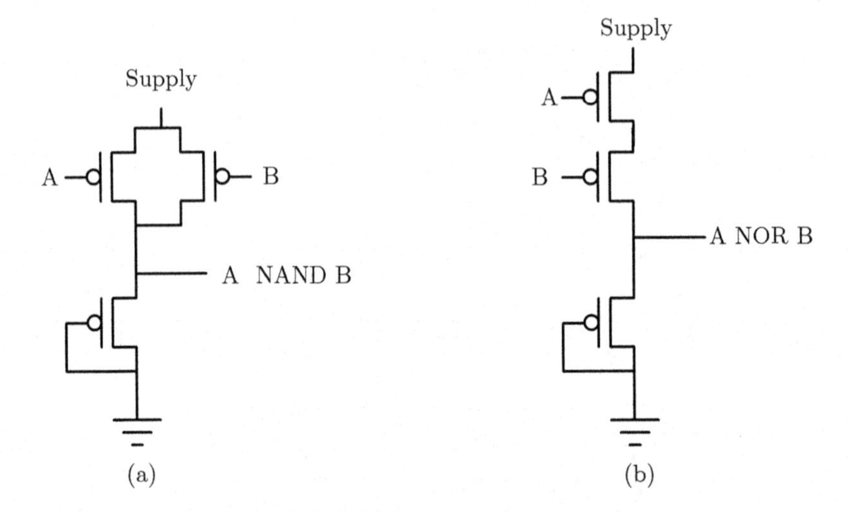

Fig. D.3. Two more sophisticated logic gates, (a) NAND and (b) NOR. The parallel transistors in the NAND gate only allow the output to fall low when both inputs are low; either high input pulls the output high. In the NOR gate the output is only pulled up when both inputs are high. Any appropriate load for the threshold voltage can be used, but a saturated load is shown for both circuits. Multiple input NAND and NOR gates are also in principle possible. Because the stacked transistor geometry limits the output swing by at least V_T over that of the NAND gate, only NAND gates are typically encountered in the literature.

latory response. The challenge with oscillators that are too small, however, is that the parasitic capacitance serves as a low pass filter which can reduce the loop gain at the frequencies needed and prevent spontaneous oscillation. Oscillators which have more inverter elements can ring with lower frequency signals and can overcome this problem. The frequency response can be easily diagnosed by looking at a driven inverter chain or extrapolating from the frequency characteristic of a single inverter.

The measurement of the ring oscillator should be performed carefully so as to avoid loading the ring and suppressing the oscillation. One approach is to add an extra inverter to buffer the output signal–this adds only one inverter's worth of capacitave load to the chain. A buffered active probe can also be used for measurement. The IEEE 1620 standard has an appendix which further discusses good practices for using ring oscillators as a transistor speed benchmark.

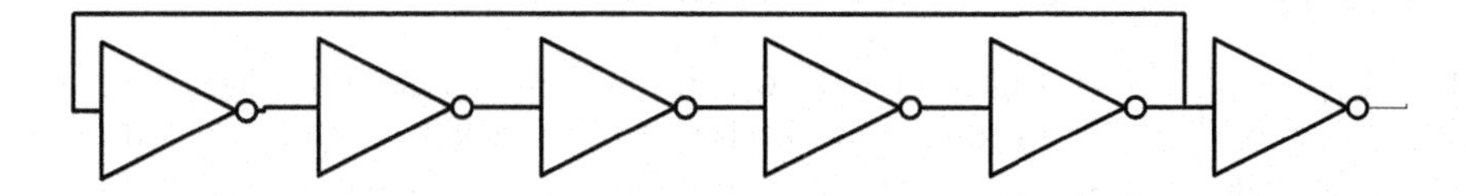

Fig. D.4. A logical diagram showing a ring oscillator. An odd number of inverters is connected, forming a metastable logic circuit if the chain is long enough and the parasitic capacitances are low relative to the drive currents of the transistors.

D.4 SRAM

An alternative application of coupled inverters is to place an even number (say, 2) together and create a bistable logic reprogrammable memory circuit. An example SRAM formed in this way is shown in Fig. D.5 [145].

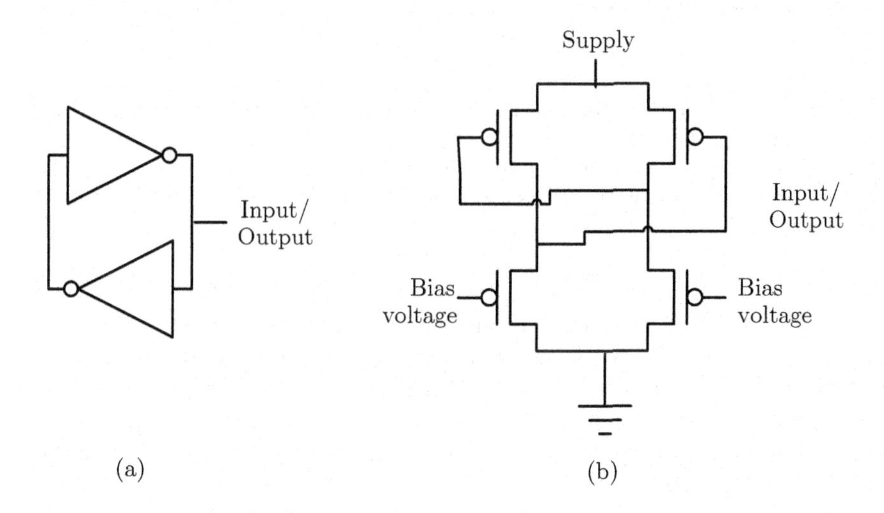

Fig. D.5. A static memory element formed by coupling two inverters together into a bistable arrangement (based on [145] and related references).

The cross-coupled inverters can be reprogrammed by driving the output with a voltage source which is stronger than the inverters. In [145], the authors further increase the noise margin of the inverters used by using a back gate to adjust the threshold voltage of the devices.

D.5 Flip flops and code generators

Flip-flops are state storage elements which have provisions for programming and, in many cases, clocking. One architecture which is useful for making clock dividers is known as the T flip-flop, whose design is shown in Fig. D.6. The T flip flop stores state when the input is steady, but inverts its output state on a rising input edge.

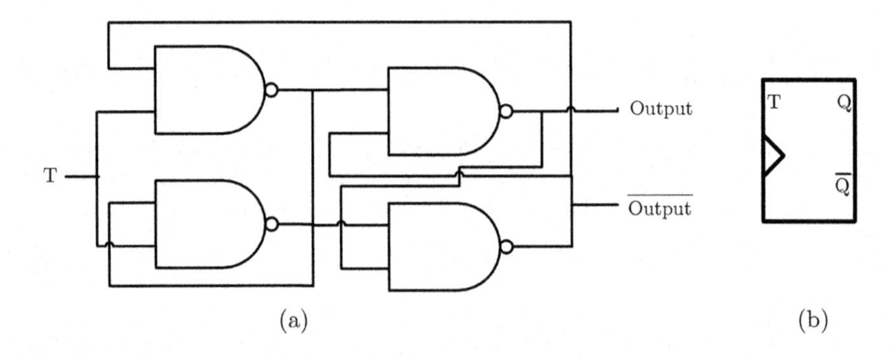

Fig. D.6. (a) A simple T-flip flop implemented using NAND gates (b) the symbol for a T flip-flop. Q is the output, $\bar{Q}$ is the inverse of the output. When a rising clock edge is applied to T, the state switches from on to off or vice versa. More sophisticated latches may be necessary to prevent race conditions depending on the transistors used.

Because the device transitions only every other transition edge, the T flip-flop serves as a divide by two counter which can be cascaded to form a power of 2 counter. Several OFET-based RFID tags have been designed which use code generators based on clock dividers to produce an arbitrary sequence that is transmitted out of the device [61] [74]. Fig. D.7 shows an example with four cascaded T flip-flops.

As a clock is applied to the input, this cascade produces all of the binary numbers between 0 and 15. Because it is positive edge triggered it counts backwards and then cycles back to 1111 to repeat again (1111, 1110, 1101, ... , 0001, 0000, 1111). This binary output sequence can then be sent to a decoder, which is a logic circuit formed by a cascade of inverters and NAND gates which produces an output based on the input. This logic cascade can produce the sequence desired from the code generator. An example canonical form is shown in Fig. D.8 designed using only inverters and NAND gates. The first stage of multi-input NAND gates, which can be formed either as a single unit or using a cascade of smaller gates, selects the combinations which produce a logical 1 output. This formulation is logically equivalent to the canonical sum of products arrangement, but replaces the OR summing

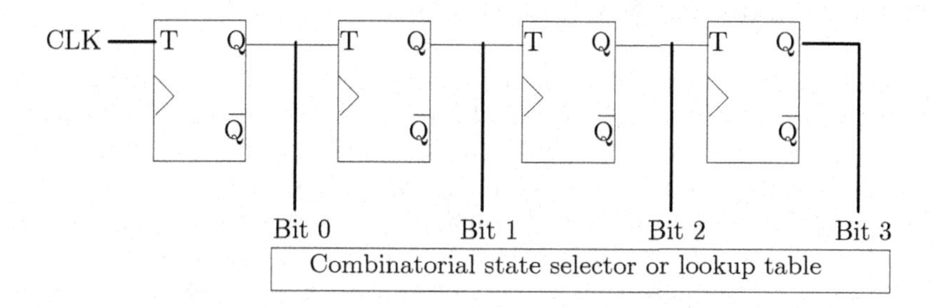

Fig. D.7.

element with a NAND element and inverted inputs from the NAND selector gates.

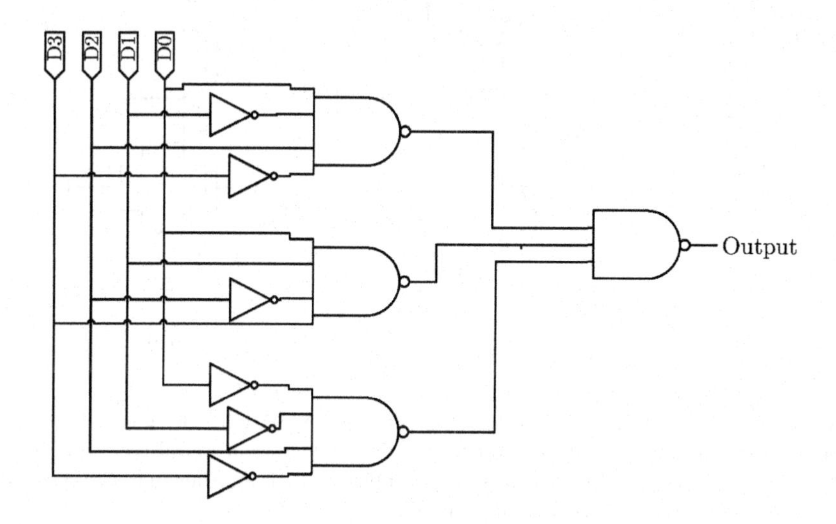

Fig. D.8. Asimple decoder made using only inverters and NAND gates which can interface with the clock in figure Fig. D.7 to form a code generator. This arrangement will produce a 1 output for inputs of 0101, 1011, and 0100; and a 0 output otherwise.

All other flip-flops can also be implemented using the NAND gate as the building block. These state retention elements can be used to make memories and more complicated digital circuitry.

D.6 Complementary OFET circuits

It is certainly possible to fabricate both complementary OFET circuits [146] or mixed complementary amorphous silicon NFET and OFET PFET circuits to obtain thin film CMOS logic [147]. In these cases significantly higher gains are possible and static power consumption can be reduced. Circuit design in these cases is straightforward; the primary concern is balancing the dimensions of the devices to maximize the noise margin and achieve the desired switching speed. Because devices must be stacked in either NAND or NOR configurations, making 2 or more input logic elements gates using CMOS logic in OFET technologies can be challenging.

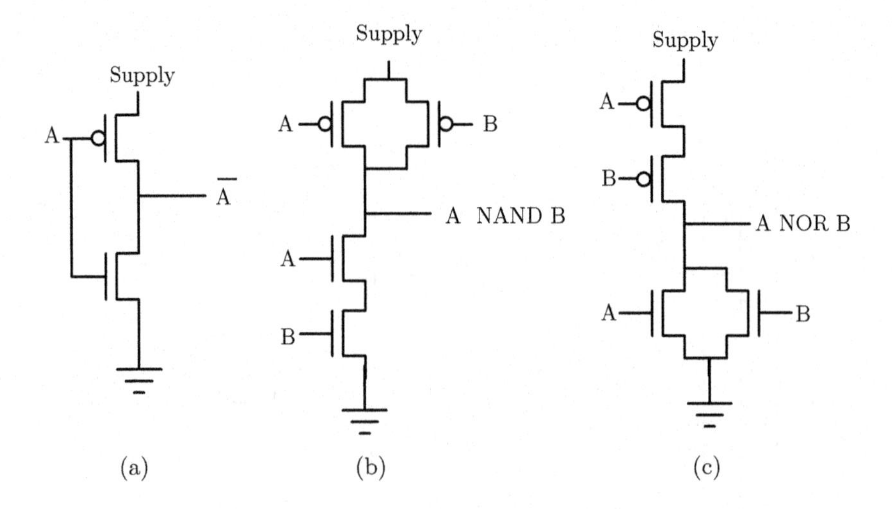

Fig. D.9. (a) Inverter, (b) NAND gate, and (c) NOR gate designs using complementary transistor elements. Complementary devices can be fabricated using n and p-type OFETs, or by combining p-type OFETs with another material, such as amorphous silicon, which produces NFETs.

References

1. F. Ebisawa, T. Kurokawa, and S. Nara. Electrical properties of polyacetylene/polysiloxane interface. *Journal of Applied Physics*, 54(6):3255–3259, 1983.
2. A. Tsumura, H. Koezuka, and T. Ando. Macromolecular electronic device: Field-effect transistor with a polythiophene thin film. *Applied Physics Letters*, 49(18):1210–1212, 1986.
3. C. W. Tang. Two-layer organic photovoltaic cell. *Applied Physics Letters*, 48(2):183–185, 1986.
4. C. W. Tang and S. A. VanSlyke. Organic electroluminescent diodes. *Applied Physics Letters*, 51(12):913–915, 1987.
5. K. Y. Jen, G. G. Miller, and R. L. Elsenbaumer. Highly Conducting, Soluble, and Environmentally-Stable Poly(3-Alkylthiophenes). *Journal of the Chemical Society-Chemical Communications*, (17):1346–1347, Sep 1 1986.
6. A. Assadi, C. Svensson, M. Willander, and O. Inganas. Field-effect mobility of poly(3-hexylthiophene). *Applied Physics Letters*, 53(3):195–197, 1988.
7. M. Pope and CE Swenberg. *Electronic processes in organic crystals and polymers.* Oxford University Press, New York, 1999.
8. PW Atkins and RS Friedman. *Molecular Quantum Mechanics.* Oxford University Press, Oxford, third edition, 1997.
9. R. Saito, G. Dresselhaus, and M. S. Dresselhaus. *Physical Properties of Carbon Nanotubes.* Imperial College Press, 1998.
10. Ioannis Kymissis. *Field Emission from Organic Materials.* PhD thesis, Massachusetts Institute of Technology, 2003.
11. Charles W. Scherr. Free-Electron Network Model for Conjugated Systems. II. Numerical Calculations. *The Journal of Chemical Physics*, 21(9):1582–1596, 1953.
12. TA Skotheim, RL Elsenbaumer, and JR Reynolds. *Handbook of conducting Polymers.* Marcel Dekker, New York, 1998.
13. Lay-Lay Chua, Jana Zaumseil, Jui-Fen Chang, Eric C. W. Ou, Peter K. H. Ho, Henning Sirringhaus, and Richard H. Friend. General observation of n-type field-effect behaviour in organic semiconductors. *Nature*, 434(7030):194–199, 2005.
14. Takashi Minakata, Masaru Ozaki, and Hideaki Imai. Conducting thin films of pentacene doped with alkaline metals. *Journal of Applied Physics*, 74(2):1079–1082, 1993.

15. Yasumitsu Matsuo, Sachio Sasaki, and Seiichiro Ikehata. Stage structure and electrical properties of rubidium-doped pentacene. *Physics Letters A*, 321(1):62–66, 2004.
16. C. K. Chan, A. Kahn, Q. Zhang, S. Barlow, and S. R. Marder. Incorporation of cobaltocene as an n-dopant in organic molecular films. *Journal of Applied Physics*, 102(1):–, Jul 1 2007.
17. C. D. Dimitrakopoulos and P. R. L. Malenfant. Organic thin film transistors for large area electronics. *Advanced Materials*, 14(2):99–+, Jan 16 2002.
18. T. Yamamoto, K. Sanechika, and A. Yamamoto. Preparation of Thermostable and Electric-Conducting Poly(2,5-Thienylene). *Journal of Polymer Science Part C-Polymer Letters*, 18(1):9–12, 1980.
19. T. A. Chen, X. M. Wu, and R. D. Rieke. Regiocontrolled Synthesis of Poly(3-Alkylthiophenes) Mediated by Rieke Zinc - Their Characterization and Solid-State Properties. *Journal of the American Chemical Society*, 117(1):233–244, Jan 11 1995.
20. G. Barbarella, A. Bongini, and M. Zambianchi. Regiochemistry and Conformation of Poly(3-Hexylthiophene) Via the Synthesis and the Spectroscopic Characterization of the Model Configurational Triads. *Macromolecules*, 27(11):3039–3045, May 23 1994.
21. B. S. Ong, Y. Wu, P. Liu, and S. Gardner. High-Performance Semiconducting Polythiophenes for Organic Thin-Film Transistors. *Journal of the American Chemical Society*, 126(11):3378–3379, 2004.
22. I. Mcculloch, M. Heeney, C. Bailey, K. Genevicius, I. Macdonald, M. Shkunov, D. Sparrowe, S. Tierney, R. Wagner, W. M. Zhang, M. L. Chabinyc, R. J. Kline, M. D. Mcgehee, and M. F. Toney. Liquid-crystalline semiconducting polymers with high charge-carrier mobility. *Nature Materials*, 5(4):328–333, Apr 2006.
23. Michael Inbasekaran, Weishi Wu, and Edmund P. Woo. Process for preparing conjugated polymers. *United States Patent 5777070*, 1998.
24. T. N. Jackson D. L. Eaton J. E. Anthony C.D. Sheraw. Functionalized Pentacene Active Layer Organic Thin-Film Transistors. *Advanced Materials*, 15(23):2009–2011, 2003.
25. J. E. Anthony. Functionalized Acenes and Heteroacenes for Organic Electronics. *Chemical Reviews*, 106(12):5028–5048, 2006.
26. A. Afzali, C. D. Dimitrakopoulos, and T. L. Breen. High-Performance, Solution-Processed Organic Thin Film Transistors from a Novel Pentacene Precursor. *Journal of the American Chemical Society*, 124(30):8812–8813, 2002.
27. M. L. Tang, T. Okamoto, and Z. Bao. High-Performance Organic Semiconductors: Asymmetric Linear Acenes Containing Sulphur. *Journal of the American Chemical Society*, 128(50):16002–16003, 2006.
28. H. Meng, Z. Bao, A. J. Lovinger, B. C. Wang, and A. M. Mujsce. High Field-Effect Mobility Oligofluorene Derivatives with High Environmental Stability. *Journal of the American Chemical Society*, 123(37):9214–9215, 2001.
29. F. Garnier, A. Yassar, R. Hajlaoui, G. Horowitz, F. Deloffre, B. Servet, S. Ries, and P. Alnot. Molecular engineering of organic semiconductors: design of self-assembly properties in conjugated thiophene oligomers. *Journal of the American Chemical Society*, 115(19):8716–8721, 1993.
30. Z. Bao, A. J. Lovinger, and J. Brown. New Air-Stable n-Channel Organic Thin Film Transistors. *Journal of the American Chemical Society*, 120(1):207–208, 1998.

31. Y. Sakamoto, T. Suzuki, M. Kobayashi, Y. Gao, Y. Fukai, Y. Inoue, F. Sato, and S. Tokito. Perfluoropentacene: High-Performance p-n Junctions and Complementary Circuits with Pentacene. *Journal of the American Chemical Society*, 126(26):8138–8140, 2004.
32. S. Ando, R. Murakami, J. i Nishida, H. Tada, Y. Inoue, S. Tokito, and Y. Yamashita. n-Type Organic Field-Effect Transistors with Very High Electron Mobility Based on Thiazole Oligomers with Trifluoromethylphenyl Groups. *Journal of the American Chemical Society*, 127(43):14996–14997, 2005.
33. R. C. Haddon, A. S. Perel, R. C. Morris, T. T. M. Palstra, A. F. Hebard, and R. M. Fleming. C60 thin film transistors. *Applied Physics Letters*, 67:121, 1995.
34. C. R. Newman, C. D. Frisbie, D. A. daSilvaFilho, J. L. Bredas, P. C. Ewbank, and K. R. Mann. Introduction to Organic Thin Film Transistors and Design of n-Channel Organic Semiconductors. *Chemistry of Materials*, 16(23):4436–4451, 2004.
35. D. M. de Leeuw, M. M. J. Simenon, A. R. Brown, and R. E. F. Einerhand. Stability of n-type doped conducting polymers and consequences for polymeric microelectronic devices. *Synthetic Metals*, 87(1):53–59, 1997.
36. T. Ashimine, T. Yasuda, M. Saito, H. Nakamura, and T. Tsutsui. Air Stability of p-Channel Organic Field-Effect Transistors Based on Oligo-p-phenylenevinylene Derivatives. *Japanese Journal of Applied Physics*, 47(3):1760–1762, 2008.
37. T.D. Anthopoulos, F. B Kooistra, H. J Wondergem, D. Kronholm, J. C Hummelen, and D. M deLeeuw. Air-Stable n-Channel Organic Transistors Based on a Soluble C84 Fullerene Derivative. *Advanced Materials*, 18(13):1679–1684, 2006.
38. Calvin K. Chan, Fabrice Amy, Qing Zhang, Stephen Barlow, Seth Marder, and Antoine Kahn. N-type doping of an electron-transport material by controlled gas-phase incorporation of cobaltocene. *Chemical Physics Letters*, 431(1-3):67–71, 2006.
39. A. Vollmer, H. Weiss, S. Rentenberger, I. Salzmann, J. P. Rabe, and N. Koch. The interaction of oxygen and ozone with pentacene. *Surface Science*, 600(18):4004–4007, 2006.
40. T. W. Kelley, D. V. Muyres, P. F. Baude, T. P. Smith, and T. D. Jones. High performance organic thin film transistors. In *Proceedings of the Material Research Society*, volume 771, page 169179, 2003.
41. Martin Knudsen. Die Gesetze der Molekularstrmung und der inneren Reibungsstrmung der Gase durch Rhren. *Annalen der Physik*, 333(1):75–130, 1909.
42. R. Feres and G. Yablonsky. Knudsen's cosine law and random billiards. *Chemical Engineering Science*, 59(7):1541–1556, 2004.
43. T. J. Mattord, V. P. Kesan, D. P. Neikirk, and B. G. Streetman. A Single-Filament Effusion Cell with Reduced Thermal-Gradient for Molecular-Beam Epitaxy. *Journal of Vacuum Science and Technology B*, 7(2):214–216, Mar-Apr 1989.
44. Yen-Yi Lin, D. I. Gundlach, S. F. Nelson, and T. N. Jackson. Pentacene-based organic thin-film transistors Pentacene-based organic thin-film transistors. *Electron Devices, IEEE Transactions on*, 44(8):1325–1331, 1997.

45. V. Podzorov, S. E. Sysoev, E. Loginova, V. M. Pudalov, and M. E. Gershenson. Single-crystal organic field effect transistors with the hole mobility 8 cm2/V s. *Applied Physics Letters*, 83(17):3504–3506, 2003.
46. A. R. McGhie, A. F. Garito, and A. J. Heeger. A gradient sublimer for purification and crystal growth of organic donor and acceptor molecules. *Journal of Crystal Growth*, 22(4):295–297, 1974.
47. O. Ostroverkhova, D. G. Cooke, F. A. Hegmann, J. E. Anthony, V. Podzorov, M. E. Gershenson, O. D. Jurchescu, and T. T. M. Palstra. Ultrafast carrier dynamics in pentacene, functionalized pentacene, tetracene, and rubrene single crystals. *Applied Physics Letters*, 88(16):162101–3, 2006.
48. Max Shtein, Jonathan Mapel, Jay B. Benziger, and Stephen R. Forrest. Effects of film morphology and gate dielectric surface preparation on the electrical characteristics of organic-vapor-phase-deposited pentacene thin-film transistors. *Applied Physics Letters*, 81(2):268–270, 2002.
49. R. Parashkov, E. Becker, T. Riedl, H. H. Johannes, and W. Kowalsky. Large Area Electronics Using Printing Methods. *Proceedings of the IEEE*, 93(7):1321–1329, 2005.
50. K. K. S. Lau, H. G. Pryce Lewis, S. J. Limb, M. C. Kwan, and K. K. Gleason. Hot-wire chemical vapor deposition (HWCVD) of fluorocarbon and organosilicon thin films. *Thin Solid Films*, 395(1-2):288–291, 2001.
51. John P. Lock, Jodie L. Lutkenhaus, Nicole S. Zacharia, Sung Gap Im, Paula T. Hammond, and Karen K. Gleason. Electrochemical investigation of PEDOT films deposited via CVD for electrochromic applications. *Synthetic Metals*, 157(22-23):894–898, 2007.
52. M. Szwarc. The C[Single Bond]H Bond Energy in Toluene and Xylenes. *The Journal of Chemical Physics*, 16(2):128–136, 1948.
53. William F. Gorham. A New, General Synthetic Method for the Preparation of Linear Poly-¡I¿p¡/I¿-xylylenes. *Journal of Polymer Science Part A-1 Polymer Chemistry*, 4(12):3027–3039, 1966.
54. Annie Wang, Ioannis Kymissis, Vladimir Bulovic, and Akintunde I. Akinwande. Tunable threshold voltage and flatband voltage in pentacene field effect transistors. *Applied Physics Letters*, 89(11):112109–3, 2006.
55. D. J. Gundlach, T. N. Jackson, D. G. Schlom, and S. F. Nelson. Solvent-induced phase transition in thermally evaporated pentacene films. *Applied Physics Letters*, 74(22):3302–3304, 1999.
56. C. D. Dimitrakopoulos, A. R. Brown, and A. Pomp. Molecular beam deposited thin films of pentacene for organic field effect transistor applications. *Journal of Applied Physics*, 80(4):2501–2508, 1996.
57. M. G. Kane, J. Campi, M. S. Hammond, F. P. Cuomo, B. Greening, C. D. Sheraw, J. A. Nichols, D. J. Gundlach, J. R. Huang, C. C. Kuo, L. A. Jia L. Jia, H. Klauk, and T. N. Jackson. Analog and digital circuits using organic thin-film transistors on polyester substrates. *Electron Device Letters, IEEE*, 21(11):534–536, 2000.
58. Ioannis Kymissis, Christos D. Dimitrakopoulos, and Sampath Purushothaman. Patterning pentacene organic thin film transistors. *Journal of Vacuum Science and Technology B: Microelectronics and Nanometer Structures*, 20(3):956–959, 2002.
59. Ha Soo Hwang, Alexander Zakhidov, Jin-Kyun Lee, John A. DeFranco, Hon Hang Fong, George G. Malliaras, and Christopher K. Ober. Photolithographic Patterning in Supercritical Carbon Dioxide: Application to Patterned

Light-emitting Devices. In *Flexible Electronics and Displays Conference and Exhibition, 2008*, pages 1–4, 2008.

60. K. P. Weidkamp, A. Afzali, R. M. Tromp, and R. J. Hamers. A Photopatternable Pentacene Precursor for Use in Organic Thin-Film Transistors. *Journal of the American Chemical Society*, 126(40):12740–12741, 2004.
61. C. J. Drury, C. M. J. Mutsaers, C. M. Hart, M. Matters, and D. M. de Leeuw. Low-cost all-polymer integrated circuits. *Applied Physics Letters*, 73(1):108–110, 1998.
62. K. R. Williams, K. R. Williams, and R. S. Muller. Etch rates for micromachining processing Etch rates for micromachining processing. *Microelectromechanical Systems, Journal of*, 5(4):256–269, 1996.
63. K. R. Williams, K. R. Williams, K. Gupta, and M. Wasilik. Etch rates for micromachining processing-Part II Etch rates for micromachining processing-Part II. *Microelectromechanical Systems, Journal of*, 12(6):761–778, 2003.
64. Soeren Steudel, Kris Myny, Stijn De Vusser, Jan Genoe, and Paul Heremans. Patterning of organic thin film transistors by oxygen plasma etch. *Applied Physics Letters*, 89(18):183503–3, 2006.
65. D. J. Gundlach, D. J. Gundlach, Y. Y. Lin, T. N. Jackson, S. F. A. Nelson S. F. Nelson, and D. G. A. Schlom D. G. Schlom. Pentacene organic thin-film transistors-molecular ordering and mobility Pentacene organic thin-film transistors-molecular ordering and mobility. *Electron Device Letters, IEEE*, 18(3):87–89, 1997.
66. B. N. Park, Soonjoo Seo, and Paul G. Evans. Channel formation in single-monolayer pentacene thin film transistors. *Journal of Physics D: Applied Physics*, 40(11):3506–3511, 2007.
67. Franco Dinelli, Mauro Murgia, Pablo Levy, Massimiliano Cavallini, Fabio Biscarini, and Dago M. de Leeuw. Spatially Correlated Charge Transport in Organic Thin Film Transistors. *Physical Review Letters*, 92(11):116802, 2004.
68. A. I. Wang. *Threshold voltage in pentacene field effect transistors with parylene dielectric.* PhD thesis, Massachusetts Institute of Technology, 2004.
69. Oana D. Jurchescu, Jacob Baas, and Thomas T. M. Palstra. Electronic transport properties of pentacene single crystals upon exposure to air. *Applied Physics Letters*, 87(5):052102–3, 2005.
70. G. Gu, M. G. Kane, J. E. Doty, and A. H. Firester. Electron traps and hysteresis in pentacene-based organic thin-film transistors. *Applied Physics Letters*, 87(24):–, Dec 12 2005.
71. Rongbin Ye, Mamoru Baba, Kazunori Suzuki, Yoshiyuki Ohishi, and Kunio Mori. Effects of O2 and H2O on electrical characteristics of pentacene thin film transistors. *Thin Solid Films*, 464-465:437–440, 2004.
72. Tsuyoshi Sekitani, Shingo Iba, Yusaku Kato, Yoshiaki Noguchi, Takao Someya, and Takayasu Sakurai. Suppression of DC bias stress-induced degradation of organic field-effect transistors using postannealing effects. *Applied Physics Letters*, 87(7):073505–3, 2005.
73. Frank-J. Meyer zu Heringdorf, M. C. Reuter, and R. M. Tromp. Growth dynamics of pentacene thin films. *Nature*, 412(6846):517–520, 2001.
74. P. F. Baude, D. A. Ender, M. A. Haase, T. W. Kelley, D. V. Muyres, and S. D. Theiss. Pentacene-based radio-frequency identification circuitry. *Applied Physics Letters*, 82(22):3964–3966, 2003.

75. Feng Bai, Todd D. Jones, Kevin M. Lewandowski, Tzu-Chen Lee, Dawn V. Muyres, and Tommie W. Kelley. Organic polymers, electronic devices, and methods. *United States Patent 7098525*, 2003.
76. I. Kymissis, I. Kymissis, A. I. Akinwande, and V. Bulovic. A lithographic process for integrated organic field-effect transistors A lithographic process for integrated organic field-effect transistors. *Display Technology, Journal of*, 1(2):289–294, 2005.
77. Huiping Jia, Erich K. Gross, Robert M. Wallace, and Bruce E. Gnade. Patterning effects on poly (3-hexylthiophene) organic thin film transistors using photolithographic processes. *Organic Electronics*, 8(1):44–50, 2007.
78. G. Gu, M. G. Kane, J. E. Doty, and A. H. Firester. An organic thin-film transistor with photolithographically patterned top contacts and active layer. In M. G. Kane, editor, *Device Research Conference, 2004. 62nd DRC. Conference Digest [Late News Papers volume included]*, pages 83–84 vol.1, 2004.
79. A. C. Arias, S. E. Ready, R. Lujan, W. S. Wong, K. E. Paul, A. Salleo, M. L. Chabinyc, R. Apte, Robert A. Street, Y. Wu, P. Liu, and B. Ong. All jet-printed polymer thin-film transistor active-matrix backplanes. *Applied Physics Letters*, 85(15):3304–3306, 2004.
80. H. Klauk, D. J. Gundlach, J. A. Nichols, and T. N. Jackson. Pentacene organic thin-film transistors for circuit and display applications. *Electron Devices, IEEE Transactions on*, 46(6):1258–1263, 1999.
81. Herbert B. Michaelson. The work function of the elements and its periodicity. *Journal of Applied Physics*, 48(11):4729–4733, 1977.
82. F. Amy, C. Chan, and A. Kahn. Polarization at the gold/pentacene interface. *Organic Electronics*, 6(2):85–91, 2005.
83. J. E. Lyon, A. J. Cascio, M. M. Beerbom, R. Schlaf, Y. Zhu, and S. A. Jenekhe. Photoemission study of the poly(3-hexylthiophene)/Au interface. *Applied Physics Letters*, 88(22):222109–3, 2006.
84. Jeong-M. Choi, D. K. Hwang, Jae Hoon Kim, and Seongil Im. Transparent thin-film transistors with pentacene channel, AlO[sub x] gate, and NiO[sub x] electrodes. *Applied Physics Letters*, 86(12):123505–3, 2005.
85. N. Koch, A. Elschner, J. Schwartz, and A. Kahn. Organic molecular films on gold versus conducting polymer: Influence of injection barrier height and morphology on current–voltage characteristics. *Applied Physics Letters*, 82(14):2281–2283, 2003.
86. M. Lefenfeld, G. Blanchet, and J. A. Rogers. High-Performance Contacts in Plastic Transistors and Logic Gates That Use Printed Electrodes of DNNSA-PANI Doped with Single-Walled Carbon Nanotubes. *Advanced Materials*, 15(14):1188–1191, 2003.
87. I. Kymissis, C. D. Dimitrakopoulos, and S. Purushothaman. High-performance bottom electrode organic thin-film transistors. *Electron Devices, IEEE Transactions on*, 48(6):1060–1064, 2001.
88. D. J. Gundlach, J. E. Royer, S. K. Park, S. Subramanian, O. D. Jurchescu, B. H. Hamadani, A. J. Moad, R. J. Kline, L. C. Teague, O. Kirillov, C. A. Richter, J. G. Kushmerick, L. J. Richter, S. R. Parkin, T. N. Jackson, and J. E. Anthony. Contact-induced crystallinity for high-performance soluble acene-based transistors and circuits. *Nat Mater*, 7(3):216–221, 2008.
89. Sung Hwan Kim, Hye Young Choi, and Jin Jang. Effect of source/drain undercut on the performance of pentacene thin-film transistors on plastic. *Applied Physics Letters*, 85(19):4514–4516, 2004.

90. A. L. Briseno, S. C. B. Mannsfeld, M. M. Ling, S. H. Liu, R. J. Tseng, C. Reese, M. E. Roberts, Y. Yang, F. Wudl, and Z. N. Bao. Patterning organic single-crystal transistor arrays. *Nature*, 444(7121):913–917, Dec 14 2006.
91. D. J. Gundlach, Jia Li Li, and T. N. Jackson. Pentacene TFT with improved linear region characteristics using chemically modified source and drain electrodes. *Electron Device Letters, IEEE*, 22(12):571–573, 2001.
92. S. Young Park, Young H. Noh, and Hong H. Lee. Introduction of an interlayer between metal and semiconductor for organic thin-film transistors. *Applied Physics Letters*, 88(11):113503–3, 2006.
93. Tsuyoshi Sekitani, Yoshiaki Noguchi, Ute Zschieschang, Hagen Klauk, and Takao Someya. Organic transistors manufactured using inkjet technology with subfemtoliter accuracy. *Proceedings of the National Academy of Sciences*, 105(13):4976–4980, 2008.
94. G. Gu, M. G. Kane, and S. C. Mau. Reversible memory effects and acceptor states in pentacene-based organic thin-film transistors. *Journal of Applied Physics*, 101(1):–, Jan 1 2007.
95. H. Klauk and U. Zschieschang. Low-Voltage, Low-Power Organic Complementary Circuits with Self-Assembled Monolayer Gate Dielectric. In *Device Research Conference, 2006 64th*, pages 213–214, 2006.
96. K. P. Pernstich, C. Goldmann, C. Krellner, D. Oberhoff, D. J. Gundlach, and B. Batlogg. Shifted transfer characteristics of organic thin film and single crystal FETs. *Synthetic Metals*, 146(3):325–328, 2004.
97. T. Sekitani, Y. Nogitchi, S. Nakano, K. Zaitsu, Y. Kato, M. Takamiya, T. Sakurai, and T. Someya. Communication sheets using printed organic nonvolatile memories. In *Electron Devices Meeting, 2007. IEDM 2007. IEEE International*, pages 221–224, 2007.
98. F. A. Yildirim, C. Ucurum, R. R. Schliewe, W. Bauhofer, R. M. Meixner, H. Goebel, and W. Krautschneider. Spin-cast composite gate insulation for low driving voltages and memory effect in organic field-effect transistors. *Applied Physics Letters*, 90(8):083501–3, 2007.
99. T. Someya, H. Kawaguchi, and T. Sakurai. Cut-and-paste organic FET customized ICs for application to artificial skin. In *Solid-State Circuits Conference, 2004. Digest of Technical Papers. ISSCC. 2004 IEEE International*, pages 288–529 Vol.1, 2004.
100. M. Scharnberg, V. Zaporojtchenko, R. Adelung, F. Faupel, C. Pannemann, T. Diekmann, and U. Hilleringmann. Tuning the threshold voltage of organic field-effect transistors by an electret encapsulating layer. *Applied Physics Letters*, 90(1):013501–3, 2007.
101. Tsuyoshi Sekitani, Shingo Iba, Yusaku Kato, and Takao Someya. Pentacene field-effect transistors on plastic films operating at high temperature above 100 [degree]C. *Applied Physics Letters*, 85(17):3902–3904, 2004.
102. P. E. Burrows, G. L. Graff, M. E. Gross, P. M. Martin, M. K. Shi, M. Hall, E. Mast, C. Bonham, W. Bennett, and M. B. Sullivan. Ultra barrier flexible substrates for flat panel displays. *Displays*, 22(2):65–69, 2001.
103. Graciela B. Blanchet, Yueh-Lin Loo, J. A. Rogers, F. Gao, and C. R. Fincher. Large area, high resolution, dry printing of conducting polymers for organic electronics. *Applied Physics Letters*, 82(3):463–465, 2003.
104. Sergey Lamansky, Jr Thomas R. Hoffend, Ha Le, Vivian Jones, Martin B. Wolk, and William A. Tolbert. Laser induced thermal imaging of vacuum-

coated OLED materials. In *Organic Light-Emitting Materials and Devices IX*, volume 5937, pages 593702–15, San Diego, CA, USA, 2005. SPIE.

105. Max Shtein, Peter Peumans, Jay B. Benziger, and Stephen R. Forrest. Micropatterning of small molecular weight organic semiconductor thin films using organic vapor phase deposition. *Journal of Applied Physics*, 93(7):4005–4016, 2003.
106. C. R. Kagan, T. L. Breen, and L. L. Kosbar. Patterning organic–inorganic thin-film transistors using microcontact printed templates. *Applied Physics Letters*, 79(21):3536–3538, 2001.
107. S. J. Kim, T. Ahn, M. C. Suh, C. J. Yu, D. W. Kim, and S. D. Lee. Low-leakage polymeric thin-film transistors fabricated by laser assisted lift-off technique. *Japanese Journal of Applied Physics Part 2-Letters and Express Letters*, 44(33-36):L1109–L1111, 2005.
108. Kazuhiro Kudo, Dong Xing Wang, Masaaki Iizuka, Shigekazu Kuniyoshi, and Kuniaki Tanaka. Schottky gate static induction transistor using copper phthalocyanine films. *Thin Solid Films*, 331(1-2):51–54, 1998.
109. Hagen Klauk, David J. Gundlach, Mathias Bonse, Chung-Chen Kuo, and Thomas N. Jackson. A reduced complexity process for organic thin film transistors. *Applied Physics Letters*, 76(13):1692–1694, 2000.
110. Henry S. White, Gregg P. Kittlesen, and Mark S. Wrighton. Chemical derivatization of an array of three gold microelectrodes with polypyrrole: fabrication of a molecule-based transistor. *Journal of the American Chemical Society*, 106(18):5375–5377, 1984.
111. Zheng-Tao Zhu, Jeffrey T. Mabeck, Changcheng Zhu, Nathaniel C. Cady, Carl A. Batt, and George G. Malliaras. A simple poly(3,4-ethylene dioxythiophene)/poly(styrene sulfonic acid) transistor for glucose sensing at neutral pH. *Chemical Communications*, (13):1556–1557, 2004.
112. Dieter K. Schroder. *Semiconductor Material and Device Characterization, Third Edition.* Wiley-IEEE Press, New York, 2006.
113. P. Lang M. Mottaghi H. Aubin G. Horowitz. Extracting Parameters from the Current-Voltage Characteristics of Organic Field-Effect Transistors. *Advanced Functional Materials*, 14(11):1069–1074, 2004.
114. IEEE standard test methods for the characterization of organic transistors and materials. *IEEE Std 1620-2004 (revision is underway)*, 2004.
115. K. Ryu, I. Kymissis, V. Bulovic, and C. G. Sodini. Direct extraction of mobility in pentacene OFETs using C-V and I-V measurements. *IEEE Electron Device Letters*, 26(10):716–718, Oct 2005.
116. Gilles Horowitz. Organic Semiconductors for new electronic devices. *Advanced Materials*, 2(6-7):287–292, 1990.
117. Kyungbum (kevin) Ryu. Characterization of Organic Field Effect Transistors for OLED Displays. *MS Thesis, Massacusetts Institute of Technology, Cambridge, MA*, June 2005 2005.
118. P. V. Necliudov, M. S. Shur, D. J. Gundlach, and T. N. Jackson. Modeling of organic thin film transistors of different designs. *Journal of Applied Physics*, 88(11):6594–6597, 2000.
119. Gilles Horowitz, Riadh Hajlaoui, Denis Fichou, and Ahmed El Kassmi. Gate voltage dependent mobility of oligothiophene field-effect transistors. *Journal of Applied Physics*, 85(6):3202–3206, 1999.

120. Howard E. Katz, X. Michael Hong, Ananth Dodabalapur, and Rahul Sarpeshkar. Organic field-effect transistors with polarizable gate insulators. *Journal of Applied Physics*, 91(3):1572–1576, 2002.
121. Taeho Jung, Ananth Dodabalapur, Robert Wenz, and Siddharth Mohapatra. Moisture induced surface polarization in a poly(4-vinyl phenol) dielectric in an organic thin-film transistor. *Applied Physics Letters*, 87(18):182109–3, 2005.
122. D. V. Lang, X. Chi, T. Siegrist, A. M. Sergent, and A. P. Ramirez. Bias-Dependent Generation and Quenching of Defects in Pentacene. *Physical Review Letters*, 93(7):076601, 2004.
123. Rui He, X. Chi, Aron Pinczuk, D. V. Lang, and A. P. Ramirez. Extrinsic optical recombination in pentacene single crystals: Evidence of gap states. *Applied Physics Letters*, 87(21):211117–3, 2005.
124. G. Gu and M. G. Kane. Moisture induced electron traps and hysteresis in pentacene-based organic thin-film transistors. *Applied Physics Letters*, 92(5):–, Feb 4 2008.
125. Y. Kunugi, K. Takimiya, N. Negishi, T. Otsubo, and Y. Aso. An ambipolar organic field-effect transistor using oligothiophene incorporated with two [60]fullerenes. *Journal of Materials Chemistry*, 14(19):2840–2841, 2004.
126. Th B. Singh, S. Gunes, N. Marjanovic, N. S. Sariciftci, and R. Menon. Correlation between morphology and ambipolar transport in organic field-effect transistors. *Journal of Applied Physics*, 97(11):114508–5, 2005.
127. C. D. Dimitrakopoulos, S. Purushothaman, J. Kymissis, A. Callegari, and J. M. Shaw. Low-Voltage Organic Transistors on Plastic Comprising High-Dielectric Constant Gate Insulators. *Science*, 283(5403):822–824, February 5, 1999 1999.
128. Lawrence Dunn, Debarshi Basu, Liang Wang, and Ananth Dodabalapur. Organic field effect transistor mobility from transient response analysis. *Applied Physics Letters*, 88(6):063507–3, 2006.
129. P. M. Alt and P. Pleshko. Scanning limitations of liquid-crystal displays. *Electron Devices, IEEE Transactions on*, 21(2):146–155, 1974.
130. C. D. Sheraw, L. Zhou, J. R. Huang, D. J. Gundlach, T. N. Jackson, M. G. Kane, I. G. Hill, M. S. Hammond, J. Campi, B. K. Greening, J. Francl, and J. West. Organic thin-film transistor-driven polymer-dispersed liquid crystal displays on flexible polymeric substrates. *Applied Physics Letters*, 80(6):1088–1090, 2002.
131. John A. Rogers, Zhenan Bao, Kirk Baldwin, Ananth Dodabalapur, Brian Crone, V. R. Raju, Valerie Kuck, Howard Katz, Karl Amundson, Jay Ewing, and Paul Drzaic. Paper-like electronic displays: Large-area rubber-stamped plastic sheets of electronics and microencapsulated electrophoretic inks. *Proceedings of the National Academy of Sciences*, 98(9):4835–4840, April 24, 2001 2001.
132. Zhou Lisong, Park Sungkyu, Bai Bo, Sun Jie, Wu Sheng-Chu, T. N. Jackson, S. Nelson, D. Freeman, and Hong Yongtaek. Pentacene TFT driven AM OLED displays. *Electron Device Letters, IEEE*, 26(9):640–642, 2005.
133. Lisong Zhou, Alfred Wanga, Sheng-Chu Wu, Jie Sun, Sungkyu Park, and Thomas N. Jackson. All-organic active matrix flexible display. *Applied Physics Letters*, 88(8):083502–3, 2006.
134. Iwao Yagi, Nobukazu Hirai, Yoshihiro Miyamoto, Makoto Noda, Ayaka Imaoka, Nobuhide Yoneya, Kazumasa Nomoto, Jiro Kasahara, Akira Yumoto, and Tetsuo Urabe. A flexible full-color AMOLED display driven by OTFTs. *Journal of the Society for Information Display*, 16(1):15–20, 2008.

135. W. F. Aerts, S. Verlaak, and P. Heremans. Design of an organic pixel addressing circuit for an active-matrix OLED display. *Electron Devices, IEEE Transactions on*, 49(12):2124–2130, 2002.
136. Bahman Hekmatshoar, Alex Z. Kattamis, Kunigunde Cherenack, Sigurd Wagner, and James C. Sturm. A novel TFT-OLED integration for OLED-independent pixel programming in amorphous-Si AMOLED pixels. *Journal of the Society for Information Display*, 16(1):183–188, 2008.
137. T. Someya. Integration of organic field-effect transistors and rubbery pressure sensors for artificial skin applications. In *Electron Devices Meeting, 2003. IEDM '03 Technical Digest. IEEE International*, pages 8.4.1–8.4.4, 2003.
138. H. Kawaguchi, T. Someya, T. Sekitani, and T. Sakurai. Cut-and-paste customization of organic FET integrated circuit and its application to electronic artificial skin. *Solid-State Circuits, IEEE Journal of*, 40(1):177–185, 2005.
139. B. Stadlober, M. Zirkl, G. Leising, N. Gaar, I. Graz, S. Bauer-Gogonea, and S. Bauer. Transparent pyroelectric sensors and organic field-effect transistors with fluorinated polymers: steps towards organic infrared detectors. *Dielectrics and Electrical Insulation, IEEE Transactions on*, 13(5):1087–1092, 2006.
140. Taeksoo Ji, Mayakuthan Kathiresan, Shiny Nair, Soyoun Jung, Venkatachalam Natarajan, Rama Mohan Rao Vishnubhatla, and Vijay K. Varadan. Design and fabrication of OTFT based flexible piezoelectric sensor. In *Nanosensors, Microsensors, and Biosensors and Systems 2007*, volume 6528, pages 65281P–6, San Diego, California, USA, 2007. SPIE.
141. I. Kymissis. Integrated Circuits and Sensors using Organic Field Effect Transistors and Photodetectors. In *International Meeting on Information Display*, Incheon, South Korea, October 2008 2008.
142. T. Someya, Y. Kato, Iba Shingo, Y. Noguchi, T. Sekitani, H. Kawaguchi, and T. Sakurai. Integration of organic FETs with organic photodiodes for a large area, flexible, and lightweight sheet image scanners. *Electron Devices, IEEE Transactions on*, 52(11):2502–2511, 2005.
143. I. Nausieda, Kyungbum Ryu, I. Kymissis, A. I. Akinwande, V. Bulovic, and C. G. Sodini. An Organic Active-Matrix Imager. *Electron Devices, IEEE Transactions on*, 55(2):527–532, 2008.
144. K. Myny, S. Van Winckel, S. Steudel, P. Vicca, S. De Jonge, M. J. Beenhakkers, C. W. Sele, N. A. J. van Aerle, G. H. Gelinck, J. Genoe, and P. Heremans. An Inductively-Coupled 64b Organic RFID Tag Operating at 13.56MHz with a Data Rate of 787b/s. In *Solid-State Circuits Conference, 2008. ISSCC 2008. Digest of Technical Papers. IEEE International*, pages 290–614, 2008.
145. Y. Kato, T. Sekitani, M. Takamiya, M. Doi, K. Asaka, T. Sakurai, and T. Someya. Sheet-type Braille displays by integrating organic field-effect transistors and polymeric actuators. *Ieee Transactions on Electron Devices*, 54(2):202–209, Feb 2007.
146. Y. Y. Lin, A. Dodabalapur, R. Sarpeshkar, Z. Bao, W. Li, K. Baldwin, V. R. Raju, and H. E. Katz. Organic complementary ring oscillators. *Applied Physics Letters*, 74(18):2714–2716, 1999.
147. M. Bonse, D. B. Thomasson, H. Klauk, D. J. Gundlach, and T. N. Jackson. Integrated a-Si:H/pentacene inorganic/organic complementary circuits. In *Electron Devices Meeting, 1998. IEDM '98 Technical Digest., International*, pages 249–252, 1998.

Index

acenes, 21
alignment marks, 122
aluminum, 34
atomic layer deposition, 43

background gas, 32
baffled source, 34
bias stress, 49, 64, 88
bond pads, 123
bottom contacts, 60
buckminsterfullerene, 24

C-V measurement, 94
C60, 24
capacitance, 94, 127
carrier velocity, 81
channel, 29
chemical vapor deposition, 41
cleaning, 112
contact resistance, 86
contact treatment, 61, 62
contacts to n-type materials, 58
Corbino gate, 70
crucible evaporation, 35
crystal frustration, 60
crystal structure, 64
cutoff, 79
cyclohexene, 65

design rules, 119, 121
doping, 15, 25
drain, 29, 57

electrochemical OFET, 72
electron beam evaporation, 37
electron poor, 15
electron rich, 15
electron transporting, 14
encapsulation, 25, 66
etchin, 45
evaporation, thermal, 30

flatband voltage, 82, 84
foil boats, 32
free electron model, 5, 11

gate, 29
gate dielectric, 63, 66, 90, 113, 114
gate leakage, 90, 95

hexadecafluoro copper pthalocyanine, 24
history, 2
hole transporting, 14
HOMO, 8
HOMO-LUMO gap, 8
hybridization, 6
hysteresis, 64, 88

IEEE 1620 standard, 75
inkjet, 53

Knudsen cell, 35

large signal model, 95
laser ablation, 45
laser induced thermal imaging, 68
linear region, 78, 93

LUMO, 8

mean free path, 31
mobility, 80, 98
molecular orbitals, 8

n-type, 14

oligifluorenes, 23
oligothiophenes, 23
organic vapor phase deposition, 68
output conductance, 91
output resistance, 91

p-type, 14
parylene, 42, 51
passivation, 25
PEDOT, 26
pentacene, 22, 115
perimeter free electron model (PFEO), 11
phosphonic acids, 64
photolithography, 43, 114
pi bond, 7
Pierels instability, 10
pinch-off, 79
plasma enhanced chemical vapor depositon, 43
polaron, 12
polyaniline, 26
polyfluorenes, 20
polyimide, 116
polymers, 17
polypyrrole, 26
polystyrene, 64
polythiophene, 2, 17, 115
polyvinyl alcohol, 52
popcorn effect, 33
printing, 41
purification, 39

quad mask, 128

regioregularity, 18
resistivity, 127
resonance, 8, 11

self aligned, 113, 115
self assembled monolayers, 65, 116
self-assembled monolayers, 61
shadow masks, 44, 50
sigma bond, 7
silanes, 64
silicon device model, 75
small molecules, 20
small signal model, 95
source, 29, 57
source/drain layer, 115
stability, 25
static induction transistor, 69
steric stabilization, 25
Stranski-Krastanov growth, 46
subthreshold slope, 91
surface energy modulation, 68

tetracene, 22
thermal gradient evaporation, 38
thiols, 61, 116
threshold voltage, 49, 78, 82, 84, 98
top contacts, 60
transfer curve, 93
transfer line, 125
transistor layout, 123

van der Pauw, 127
van der Waals bonds, 29
verniers, 120

work function, 57

Continued from page ii

Routing Congestion in VLSI Circuits: Estimation and Optimization
Prashant Saxena, Rupesh S. Shelar, Sachin Sapatnekar
ISBN 978-0-387-30037-5

Ultra-Low Power Wireless Technologies for Sensor Networks
Brian Otis and Jan Rabaey
ISBN 978-0-387-30930-9

Sub-Threshold Design for Ultra Low-Power Systems
Alice Wang, Benton H. Calhoun and Anantha Chandrakasan
ISBN 978-0-387-33515-5

High Performance Energy Efficient Microprocessor Design
Vojin Oklibdzija and Ram Krishnamurthy (Eds.)
ISBN 978-0-387-28594-8

Abstraction Refinement for Large Scale Model Checking
Chao Wang, Gary D. Hachtel, and Fabio Somenzi
ISBN 978-0-387-28594-2

A Practical Introduction to PSL
Cindy Eisner and Dana Fisman
ISBN 978-0-387-35313-5

Thermal and Power Management of Integrated Systems
Arman Vassighi and Manoj Sachdev
ISBN 978-0-387-25762-4

Leakage in Nanometer CMOS Technologies
Siva G. Narendra and Anantha Chandrakasan
ISBN 978-0-387-25737-2

Statistical Analysis and Optimization for VLSI: Timing and Power
Ashish Srivastava, Dennis Sylvester, and David Blaauw
ISBN 978-0-387-26049-9